Navigation Accidents and their Causes

Navigation Accidents and their Causes

Published by The Nautical Institute
202 Lambeth Road, London SE1 7LQ, England
Tel: +44 (0)20 7928 1351   Fax: +44 (0)20 7401 2817   Web: www.nautinst.org
© The Nautical Institute 2015

All rights reserved. No part of this publication may be reproduced, stored in a retrieval system, or transmitted in any form or by any means, electronic, mechanical, photocopying, recording or otherwise, without the prior written consent of the publisher, except for quotation of brief passages in reviews.

Although great care has been taken with the writing of the book and the production of the volume, neither The Nautical Institute nor the contributors can accept any responsibility for errors and omissions or their consequences.

This book has been prepared to address the subject of navigation accidents and their causes. This should not, however, be taken to mean that this document deals comprehensively with all of the concerns that will need to be addressed or even, where a particular matter is addressed, that this document sets out the only definitive view for all situations. The opinions expressed are those of the contributors only and are not necessarily to be taken as the policies or views of any organisation with which they have any connection.

Readers of *Navigation Accidents and their Causes* are advised to make themselves aware of any applicable local, national or international legislation or administrative requirements or advice which may affect decisions taken on board.

Cover image The New Zealand Herald/newspix.co.nz
Book Editor Margaret Freeth
Typesetting and layout by Phil McAllister Design
Printed in the UK by Witherbys Ltd
ISBN 978 1 906915 32 2

THE NAUTICAL INSTITUTE

Acknowledgements

This book tackles an important question – why do navigation accidents keep happening?

The Nautical Institute would like to thank all who were involved in trying to answer it, either as authors and peer reviewers or by giving advice and support.

Special thanks are due to the following:

The Technical Editor, David Pockett Master Mariner BSc FNI, who together with the Institute's Director of Projects, David Patraiko FNI, decided the contents of the book, recruited the authors, technically reviewed their contributions and managed the peer review process.

Andrew Winbow MSc FCIArb FNI, Assistant Secretary-General/Director Maritime Safety Division, IMO, for his support.

All the authors for their generosity in sharing their knowledge, expertise and their enthusiasm for the project.

Our peer reviewers: Captain Sarabjit Butalia MSc (WMU) FNI; Captain Paul Chapman FNI; John Clandillon-Baker FNI; Captain Jeff Cowan AFNI; Captain Malcolm Goodfellow MNI; Steven Gosling MSc MNI; Captain Keith Hart BSc ExC RD MRIN FNI; Captain David Ireland; Captain Leslie R Morris BSc (Hons) FNI MRINA; Dr Andy Norris FNI; Mike Sollosi; Commodore David Squire CBE MNM FNI FCMI.

Authors would like to acknowledge the help and support of their fellow authors and peer reviewers, and the cooperation of the maritime authorities that gave permission for use of material from their accident reports.

They would also like to thank:

Captain Rory Main, Fremantle Pilots.

Colleagues at the Department of Shipping and Marine Technology at Chalmers University of Technology, Sweden.

Norman Cockcroft, lecturer and mentor for Extra Masters, and the Masters of The Ben Line Steamers who knew a thing or two about how to apply the Colregs in very busy waters.

This book builds on two books written by the late Captain Richard Cahill MBA FNI, *Collisions and their Causes* and *Strandings and their Causes*, which The Nautical Institute was pleased to publish. They proved very popular and ran to several editions.

While *Navigation Accidents and their Causes* describes the same failings that Cahill identified so clearly in his books, it also looks to the future to identify trends that may have an impact on navigational risk and suggests ways to mitigate these.

Contents

Foreword iii

Case studies iv

Introduction ix
David A Pockett

Chapter 1 Crew, manning and fatigue 1
Professor Andrew P Smith, Paul H Allen and Dr Emma J K Wadsworth

Chapter 2 The need and value of passage planning 9
Captain Nicholas Cooper

Chapter 3 Bridge resource management 17
Captain Robert Hone

Chapter 4 Positioning 25
Captain Paul Whyte

Chapter 5 A rough guide to collision avoidance 39
John Third

Chapter 6 Pilotage 49
Captain Richard Wild

Chapter 7 Under-keel clearance 57
Dr Tim Gourlay

Chapter 8 Anchoring 71
Captain Nadeem Anwar

Chapter 9 Electronics: Some friendly advice on bridge work 83
Professor Thomas Porathe

Chapter 10 Vessel Traffic Services 91
Captain Terry Hughes

Chapter 11 Learning from accidents and near misses 101
Captain Leslie R Morris

Chapter 12 Onboard training and mentoring 109
Captain André L Le Goubin

Index 115

Contributors 124

Foreword

By Koji Sekimizu
Secretary-General, International Maritime Organization

While the safety of maritime navigation continues to improve, thanks in no small part to the effective implementation of IMO measures and the effective education and training of ships' officers, accidents do, nevertheless, still frequently occur. When they do, human error is often found to be a significant causal factor. This timely publication from The Nautical Institute should provide a crucial guide for every mariner serving at sea and serve to assist in reducing collisions and groundings.

The chapters are written by an international group of authors with relevant knowledge and experience, having served as accident investigators, Master Mariners, navigation specialists and university lecturers. Key issues of concern have been addressed, including the use and misuse of the collision regulations, crew fatigue and over-reliance on electronic navigation aids.

The authors have used their experience and knowledge to look at these and other issues which have been a major cause of mistakes that have led to collisions and groundings. Previous casualties have been used to illustrate where failures have occurred and lessons which can be learned from these. The need for risk assessment in advance of a voyage is highlighted in many ways, including bridge resource management and passage planning. Situational awareness is highlighted throughout. Time has been taken to examine practical ways forward for those on the bridge to consider the risks, plan for them and then take action to avoid them. The authors have also taken a look into the future, to identify trends that may impact on navigational risk and suggest ways to mitigate them.

If we are to learn from accidents we have to consider where the best place to do so might be. Onboard training and mentoring may hold the key, and the navigation bridge is an ideal place for this to take place. At the same time, however, relevant training ashore is equally important and should be run in parallel.

The publication is written in maritime English for international mariners. Each chapter can be read individually, thus forming a valuable onboard resource. The overall message is that everyone can learn from the mistakes of others and everyone has a part to play in ensuring that training and experience are used effectively to keep vessels safe.

Case studies

There are multiple lessons to be learned from the case studies in this book and you will see several references to the same casualties in the text.

The case studies only give an outline of the casualty and we hope that you will be motivated to look at a more comprehensive account in the report produced by the relevant authorities.

Links to these can be found in the reference list below. Separately some authors have given references to their sources at the end of each chapter for further reading.

***Ever Decent*, container ship/*Norwegian Dream*, cruise ship, collision in English Channel, August 1999.**
Bahamas Maritime Authority report, www.bahamasmaritime.com
Chapter 2, page 10

***Costa Concordia*, cruise ship, grounding and capsize, Giglio Island, Italy, January 2012.**
Ministry of Infrastructures and Transports, Marine Casualties Investigative Body report on the safety technical investigation, presented by the Italian delegation to MSC 92 (12-21 June, 2013)
Chapter 2, page 12; Chapter 3, pages 19, 20; Chapter 4, page 35

***Maersk Kendal*, container ship, grounding off Singapore, September 2009.**
UK MAIB report 2/2010, www.maib.gov.uk
Chapter 3, page 21

***K-Wave*, container ship, grounding off Malaga, Spain, February 2011.**
UK MAIB report 18/2011, www.maib.gov.uk
Chapter 3, page 22

***Ficus*, oil/chemical tanker, grounding off New Providence Island, Bahamas, February 2008.**
Isle of Man Ship Registry report CA103, www.gov.im
Chapter 4, page 25

***Atlantic Blue*, products tanker, grounding Torres Strait, February 2009.**
ATSB report 262 MO-2009-001, www.atsb.gov.au
Chapter 4, page 30-31

***Oliva*, bulk carrier, grounding Nightingale Island near Tristan Da Cunha, March 2011.**
Transport Malta Marine Safety Investigation Unit report 14/2012, https://mti.gov.mt/en/Pages/MSIU/Marine-Safety-Investigation-Unit.aspx
Chapter 4, page 33

Case studies

Pride of Canterbury, passenger ferry, grounding on charted wreck near south-east coast of England, January 2008.
UK MAIB report 2/2009, www.maib.gov.uk
Chapter 4, page 34; Chapter 9, page 85

Ovit, oil/chemical tanker, grounding Dover Strait, September 2013.
UK MAIB report 24/2014, www.maib.gov.uk
Chapter four, page 34; Chapter 9, page 86; Chapter 11, page 106

Sea Diamond, sinking off Santorini, Greece, April 2007.
Official report not available.
Chapter 4, page 35

River Embley, bulk carrier, grounding Torres Strait, May 1987.
ATSB report 19 (1988), www.atsb.gov.au
Chapter 7, page 58

Queen Elizabeth 2, passenger liner, grounding Vineyard Sound, USA, August 1992.
UK MAIB report 1993, www.maib.gov.uk
Chapter 7, page 58

Sea Empress, oil tanker, grounding Milford Haven, UK, February 1996.
UK MAIB report 1997, www.maib.gov.uk
Chapter 7, page 58

Jody F Millennium, bulk carrier, grounding Gisborne, New Zealand, February 2002.
Maritime Safety Authority of New Zealand Accident Investigation report 02 2828, www.maritimenz.govt.nz
Chapter 7, pages 58, 64-65

Capella Voyager, oil tanker, grounding Marsden Point, Whangarei New Zealand, April 2003.
Maritime Safety Authority of New Zealand Accident Investigation report 03 3177, www.maritimenz.govt.nz
Chapter 7, pages 59, 63, 66

Eastern Honor, oil tanker, grounding Marsden Point, Whangarei, New Zealand, July 2003.
Referred to in Maritime Safety Authority of New Zealand Accident Investigation report 03 3177, www.maritimenz.govt.nz
Chapter 7, page 59

Desh Rakshak, oil tanker, grounding in the entrance to Port Philip, Victoria, January 2006.
ATSB report 223 (2007), www.atsb.gov.au
Chapter 7, pages 59, 63

Case studies
Navigation Accidents and their Causes

Ropax 1, ro-pax ferry, grounding Algeciras Bay, Spain, December 2008.
UK MAIB preliminary examination report February 2009, www.maib.gov.uk
Chapter 8, pages 74, 78

Pasha Bulker, bulk carrier, grounding Newcastle, NSW, Australia, June 2007.
ATSB report 243 (2007), www.atsb.gov.au
Chapter 8, pages 74, 79; Chapter 11, page 104

Willy, tanker, grounding Cornish coast, UK, January 2002.
UK MAIB report 31/2002, www.maib.gov.uk
Chapter 8, page 76

Stena Alegra, ro-pax ferry, grounding Karlskrona, Sweden, October 2013.
UK MAIB report 12/2014, www.maib.gov.uk
Chapter 8, page 76, 77

Young Lady, oil tanker, grounding off Teesport, UK, June 2007.
UK MAIB report 3/2008, www.maib.gov.uk
Chapter 8, pages 78, 79

Sleipner, high speed ferry, grounding Norway, November 1999.
Commission of Inquiry report NOU 2000:31 (in Norwegian), www.regjeringen.no
Chapter 9, page 84

Furness Melbourne, bulk carrier/***Riga II***, yacht, collision north of Bowen, Queensland, Australia, May 2012.
ATSB report 295-MO-2012-006, www.atsb.gov.au
Chapter 9, page 84

LT Cortesia, container ship, grounding Dover Strait, January 2008.
Federal Bureau of Maritime Casualty Investigation (BSU), Germany, report 01/08, www.bsu-bund.de
Chapter 9, pages 85, 86

Trans Agila, general cargo ship, grounding north of Kalmar, Sweden, November 2012.
Swedish Accident Investigation Authority report RS 2014:05,
http://www.havkom.se/assets/reports/English/RS2014_05e.pdf
Chapter 9, page 87

Godafoss, container ship, grounding off Fredrikstad, Norway, February 2011.
Accident Investigation Board of Norway report 2012/09, www.aibn.no
Chapter 9, pages 87-88

Alandsfarjan, passenger ferry, grounding approaching Mariehamn, Aland, Finland, October 2007.
Swedish Maritime Administration Report B 2008-4 (in Swedish)
Chapter 9, page 88

Tarnfjord, product tanker/*Wellamo*, ro-pax, near collision Hastholmen, Stockholm approach, August 1991.
Statens Haverikommision Rapport S 1992:1 Arende S-06/91. Summary in Swedish, www.canit.se/gms/haveri.htm
Chapter 9, page 89

Royal Majesty, passenger liner, grounding Nantucket, USA, June 1995.
National Transportation Safety Board report NTSB MAR-97-01, www.ntsb.gov
Chapter 9, page 90

Maersk Dover, ro-pax ferry/*Apollonia*, tanker/*Maersk Vancouver*, container ship, near miss Dover Strait, October 2006.
UK MAIB report 9/2007, www.maib.gov.uk
Chapter 11, page 104

Fedra, bulk carrier, grounding Gibraltar, October 2008.
Gibraltar Maritime Authority (GMA) www.gibraltarship.com
Chapter 11, pages 104-105

Full City, bulk carrier, grounding Såstein, Norway, July 2009.
Accident Investigation Board of Norway (AIBN), www.aibn.no
Chapter 11, page 105

Shen Neng 1, bulk carrier, grounding Douglas Shoal near Gladstone, Queensland, Australia, April 2010.
ATSB report 274-MO-2010-003, www.atsb.gov.au
Chapter 11, page 105

Rena, container ship, grounding Astrolabe Reef, Bay of Plenty, North Island, New Zealand, October 2011.
Transport Accident Investigation Commission (TAIC) New Zealand report 11-204, www.taic.org.nz
Chapter 11, page 105

Hyundai Dominion/*Sky Hope*, container ships, collision in East China Sea, June 2004.
UK MAIB report 17/2005, www.maib.gov.uk, following joint investigation with Hong Kong Marine Department (HKMD)
Chapter 11, page 106

CMA CGM Florida, container ship/*Chou Shan*, bulk carrier, collision in East China Sea, March 2013.
UK MAIB report 11/2014, www.maib.gov.uk
Chapter 11, page 107

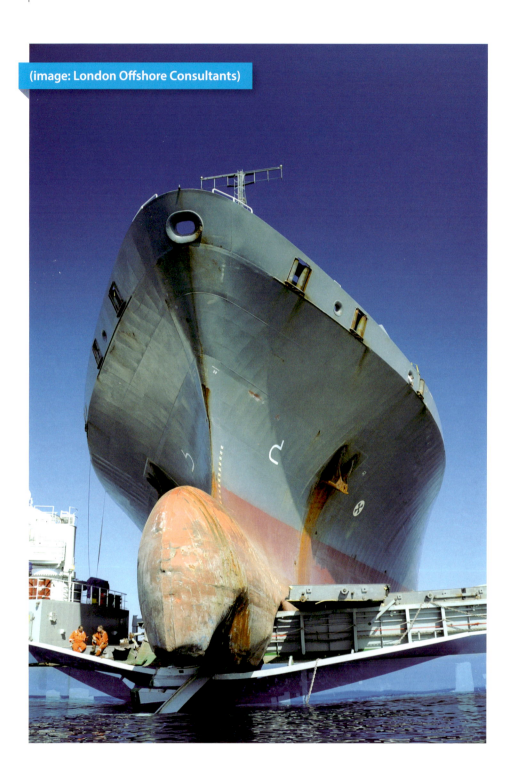
(image: London Offshore Consultants)

THE NAUTICAL INSTITUTE

Introduction

David A Pockett, Technical Editor, *Navigation Accidents and their Causes*

Despite the advanced techniques in navigation, the Safety Management System and so many other guidelines and regulations, ships still get involved in collisions and run aground, or at worst, sink. Unfortunately, human error remains a prime cause of these types of accidents. The claims statistics of one of the largest P&I Clubs, the UK P&I Club, produced in 1997 found that deck officer and pilot error accounted for over 80% of the causes of collisions. The remainder were down to equipment, mechanical or structural failure and error by those ashore. A review and analysis carried out by the American Bureau of Shipping (ABS) in 2003 entitled *ABS Review and Analysis of Accident Databases* (Clifford C Baker & Denise B McCafferty) found similar statistics for human error. It seems that statistics have not shown any marked improvements over the passage of time and human error remains a primary cause of incidents such as collisions and groundings. On the contrary, human error still gives cause for concern.

Further analysis has shown that human error in groundings and collisions has resulted from some basic failings, including a misunderstanding of the Colregs (Convention on the International Regulations for Preventing Collisions at Sea, 1972) and their application, poor bridge resource management and, in an alarming number of cases, crew fatigue.

It can be argued that it is hardly surprising there are collisions arising from human error in congested waters. Nevertheless, the counter argument has to point to collision avoidance measures which are clearly provided for in the Colregs and well supported by the navigation aids and resources on board ships today.

Collisions do result because Colregs are not followed. Rule 5 (keeping a lookout), Rule 6 (safe speed), Rule 7 (risk of collision) and Rule 8 (positive action to avoid collision), remain the key rules which seem to feature in so many collisions, and indeed groundings. Why is it that these fundamental principles of navigation and seamanship are ignored? Is it a lack of training, experience, mentoring, misunderstanding of the navigation tools provided or just plain ignorance?

Steamship Mutual P&I Club's excellent DVDs on collisions, *Collision Course* (2008), and groundings, *Groundings – Shallow Water, Deep Trouble* (2013), provide stark reminders of the consequences of human error and the best actions to avoid accidents.

Chapter 5 of this publication devotes particular attention to Colregs, which navigators would do well to heed.

An alarming number of groundings originate from ships dragging while at anchor. Again, failure to observe fundamental principles is so often the cause: an appreciation of the location and what shelter it may provide and the prevailing environmental conditions (wind, wave and current), an adequate swinging circle, an adequate scope of chain, a vigilant anchor watch to check the position at regular intervals and keep a

Introduction
Navigation Accidents and their Causes

watchful eye on the weather. All these factors and others are drummed into us at college and ought to be practised at sea. And yet ships still anchor on a lee shore, the signs of adverse weather are still ignored and a dragging anchor goes un-noticed until it is too late to do anything about it. Anchoring is addressed in Chapter 8, which focuses on many of the shortcomings in procedures which lead to incidents.

Pilot error cannot be ignored as a possible cause of accidents and statistics bear testament to this. Inadequate knowledge of the ship's characteristics, failure to discuss the un-berthing or berthing plan properly, or at all, (although the Master is implicated here too), and other factors have led to pilot-related accidents. Statistics provided by the Swedish Club in 2011 found that there was a pilot on board their insured vessels 53% of the time during a collision in congested waters. On the other hand, pilots have local knowledge and experience which should be an invaluable asset to the bridge team. Invariably they will have superior knowledge of shiphandling and manoeuvring into and out of a berth and ports, but at the same time will need full bridge resources in support of this. Chapter 6 addresses pilotage in enlightening detail.

Modern navigation techniques, with the wide array of electronic aids, are there to improve the safety of navigation. Over reliance on, or misuse of these, can lead to incidents and Chapter 9 gives examples and advice on this.

Are lessons learned from accidents? One might have thought so, but there is still a trend of the same causes repeating themselves. Chapter 11 serves as a useful reminder of how we can learn from accidents and near misses with due reference to some typical causes. And the most appropriate forum for learning? On board training and mentoring must surely be the answer and the navigation bridge the ideal venue. At the same time, relevant shore training is equally important and should be run in parallel.

Trends in the industry

The paper chart has been largely superseded by the electronic chart – ECDIS – and by 2018 all vessels over 500gt will have to use ECDIS. In other words, one might say that the paperless office has extended from ashore to ships! But can ECDIS be wholly relied upon?

Certainly not! ECDIS is also prone to human error and must be integrated in the bridge resources as a key factor for navigation. Fixing the ship's position on charts will always be a critical feature of navigation and techniques including parallel indexing, plotting the swinging circle at anchor as well as continuous monitoring will always be the primary tools for safe navigation. Chapter 4 addresses plotting. With the wide array of electronic navigational aids, there is even more emphasis on situation awareness and the need for a proper understanding of bridge resources and management.

The navigator today is seen at the control centre in the bridge surrounded by screens with displays. While the bridge window and bridge wings are far less featured, they are just as important tools for navigation as they have always been. During my many accident investigations, I have often witnessed rusted-up bridge wing gyro repeater covers indicating that they played no part in a ship's navigation. Also, collision assessments revealed that the bridge window and wings were not used to enhance a good lookout. There is certainly a strong argument for 'back to the future' and the need to rejuvenate the lost traditional skills with navigators also using their eyes beyond the wheelhouse and the comfortable chair in front of rows of control panels, displays and neon lights.

And so, with all the equipment on board ships today, there has to be the requisite training, mentoring, knowledge, skills, attentiveness and management to ensure this is used to the maximum benefit. Teamwork on board has never been more important and nor has situation awareness and the critical need to fully understand the resources at the navigator's disposal.

Add to this the increasing regulation provided from shore in the form of vessel traffic services and sea traffic management, coupled with yet more restricted areas for navigation, it is easy to see how spatial concerns will come to the fore more often in the future.

The human element

The human element plays a key role in safe voyage prosecution as it does, unfortunately, in errors that can result in collisions or groundings, minor or catastrophic. Ships' crews today are, more often than not, a mix of nationalities, languages and cultures. The need for clear understanding and fluent channels of communication has never been more important. Training and mentoring (Chapter 12) are critical ingredients in the development of efficient bridge teams, with the key objective of continuous improvement.

The plethora of electronic navigation aids poses new challenges and requires new attitudes from ships' crews. We have the human-machine interface with the navigator requiring an understanding of how systems are programmed, whether something is wrong and how to fix it. There is a complete organisational shift and different pressures on navigators today. They play more of a monitoring than controlling role insofar as navigation is concerned. New levels of competence are required to cover the technical and organisational aspects.

Monitoring the vessel's position during a voyage and having the skills required to ensure safe passage through areas of potential risk, are crucial for the navigator. Here, human intervention is critical. Old fashioned navigation techniques and principles are as important today as they were yesterday and can be used with electronic aids just as effectively, if not more so. Plotting a ship's safe course and progress is afforded particular attention in Chapter 4.

And then we have Human Element Leadership and Management (HELM). Now a requirement under Regulations II (navigating officer in charge of a watch) and III (engineer officer in charge of a watch) on Standards of Training, Certification and Watchkeeping for Seafarers (STCW), it provides the watch-keeping navigator and engineer with the knowledge and awareness of the key human factors which can influence effective resource management. The objectives of HELM are to improve safety at sea. In so doing, the emphasis is on risk assessment, situation awareness, good communications, on board training, cultural issues with the international make-up of crews today, the need for teamwork, recognition and understanding of workload and how this might be best managed to avoid problems of fatigue. Above all, management skills to address these issues to obtain the best results are afforded particular attention.

Actions and decisions taken shoreside affect those seafarers whose errors have been shown to be the cause of such a high percentage of navigation accidents. Owners, management and authorities play a role in so many ways. Investment in training, the provision of good working conditions including leave, shore/ship communications, efficient and meaningful superintendence are all key considerations. Crew manning scales set by flag states dictate watchkeeping arrangements, which in turn can have an impact on the risk of fatigue. Safe manning scales should have that mind. Do they? How can profit margin be weighed against safety and efficiency? A serious question to consider, although the answer should be obvious.

Managing risks

What better way to manage risks than to identify them, then discuss and plan to avoid them occurring. Bridge resource management (Chapter 3) is the ideal forum for this and can take the members of the bridge team, including the pilot (a temporary member of the team) and chief engineer, through the various stages of a voyage.

Before departure, a passage plan should be prepared which sets the courses, identifies the risks en route, the 'no go' areas, and directs the navigating officers to the relevant publications to be consulted or downloaded for the particular voyage. Chapter 2 addresses passage planning and the crucial part it plays in safe navigation.

The Master/pilot exchange provides the perfect opportunity for sharing of knowledge before a ship leaves the berth. The pilot can be tuned in to the ship's characteristics and the Master will have an understanding of the pilot's un-berthing plan, use of tugs or otherwise and navigation from the port. The same applies to port arrival and berthing. Chapter 6 touches upon the Master/pilot exchange and also the pilot's position within the bridge team.

Fatigue is an ever increasing problem with reduced manning scales, fast port turnarounds, long hours and constant commercial pressure to keep on schedule. Fatigue has been cited on many occasions as a cause of incidents. Consider the effect on one's mental alertness and ability to make timely decisions with a clear mind; at worst, a watchkeeper asleep on the bridge in the chair so kindly provided as a bridge comfort today whereas in yesteryear it was reserved for the pilot! Chapter 1 addresses crew manning and fatigue and gives an insight into the associated problems and measures that can be taken to avoid overwork and provide conditions that enable watchkeepers to keep alert and able to make well-informed decisions.

THE NAUTICAL INSTITUTE

Safe draught is another feature which must be foremost in a Master's mind before and during the prosecution of a voyage. So often ships have run aground during river transits, while berthing and un-berthing, when dragging anchor or even on passage steering a course into shallow water believing there to be adequate under-keel clearance. Chapter 7 provides an ideal background into this and how to manage the risks.

Vessel traffic services (VTS) are an important aid which, if used properly, enhances the safety of navigation into and out of ports and harbours, estuaries and in coastal areas with heavy traffic density such as the Dover Strait. Clear and concise channels of communication between ship and shore, with alert, competent navigators and the appropriate aids to navigate a ship safely, are crucial ingredients to manage and reduce risk. Chapter 10 provides an excellent guide to VTS and the services offered.

Predicting the future

The pace of progress has been such over the last two decades that if it is maintained, the unthinkable could well come to fruition: driverless ships, remotely controlled from a shore base; specialist sea-going personnel with not only navigation knowledge but able to program, maintain and repair electronic aids during a voyage; berthing and un-berthing by dynamic positioning control; robots able to do the tasks of crew? Reality or fantasy?

And then we have the ever improving VTS. It is suggested that the instructions from the VTS station will ease the burden of the navigator. Might it take away the responsibility of the Master as well in due course? Will it be said that Masters may not be in sole charge of the safety of their ships?

What about the bridge window, the need for human intervention and, most important, the experience which can never be surpassed? Will electronic navigation aids be fail-safe? Will there be no need for situation awareness? Can this be sensed from a shore station or base? In other words, will the salt leave the sea? Definitely not!

Ships' crews can never be a thing of the past. On board human intervention is, and will always be, a must. Electronic navigation aids can never be fail-safe and will always require an operator. Machinery requires maintenance and care, as does steelwork and ships' fabric. Moorings have to be manhandled and the presence of crew demands hotel services.

Where there will be changes is how and when the burden of responsibility might shift between a Master, navigator and shore control. It is possible to foresee a voyage being prosecuted from berth to berth with crew intervention only for seamanship and maintenance and repair duties. Navigation might well be performed without human intervention other than to ensure that the instruments have the correct input data and to provide corrective or repair measures. At the same time, however, it is hard to consider a situation whereby the Master is not in overall command and responsible for the safety of the ship.

Automated systems exist today which enable navigation from berth to berth. However, they too can go wrong in much the same way as a jumbo jet's automatic landing system. The plane's pilot is there to take control in the event of failure as too will be the Master and crew on board a ship for the same reason.

Driverless ships and run from ashore? Is it really possible? Can the con be transferred entirely? Will we say good-bye to the bridge window? Conceivable perhaps but a daunting prospect. Take away human intervention on the spot and one loses the core experience and expertise which is built up over many years in the maritime environment. Situation awareness too can best be provided from the source and not remotely.

Navigation aids are only as good as the user and will be in constant need of an alert observer who understands the input and output, can assess the data provided and identify faults. The navigator will still play an important role but the job specification will be wider and more sophisticated than before.

Spacial issues too will become ever more of a challenge. The continued exploration for hydrocarbons offshore and implementation of renewable energy systems do not come without an impact on navigation, particularly in coastal areas. New exclusive economic zones, reduced sea room, greater regulatory measures and the need for yet tighter control all suggest a leaning to a 'Big Brother' approach in the future.

The navigator navigating or being navigated, or perhaps active to passive navigation? An interesting concept. The many issues featured in this publication will hopefully give much food for thought.

References

UK P&I Club. Analysis of Major Claims 1997, www.ukpandi.com

Baker, Clifford C and McCafferty Denise B. ABS Review and Analysis of Accident Databases, 2003

Swedish Club. *Collisions and Groundings*, 2011, www.swedishclub.com

Chapter 1

Crew, manning and fatigue

By Professor Andrew P Smith, Paul H Allen and Dr Emma J K Wadsworth

Fatigue is a leading cause, or contributing factor in, maritime accidents, collisions and groundings. This chapter considers the problem and aims to provide a better understanding of the signs, risks and consequences of fatigue and how it can be addressed.

Fatigue is often defined as excessive tiredness or exhaustion. Both physical and mental fatigue can occur and a simple way of conceptualising fatigue is to consider a *fatigue process*. This starts with risk factors for fatigue (such as long working hours and sleep loss), then considers perceptions of fatigue and, finally, fatigue outcomes (eg changes in mood; inefficient performance; accidents and injury).

Occupational fatigue has been widely studied onshore and in other areas of transport; results suggest that over 20% of the working population reports feeling very fatigued at work. It would appear that work-related fatigue is an everyday experience which can have devastating personal, industrial and environmental consequences. Work-related fatigue accidents are estimated to cost £240 million a year in the UK alone.

Fatigue has also been shown to impair performance, with effects often being as large as those seen with moderate doses of alcohol. Driver fatigue is a major cause of road accidents and fatigue is an established risk factor in the aviation industry. Fatigue has been identified as a root cause in major accidents such as the *Herald of Free Enterprise* disaster.

Although there is relatively little information on recognising and managing fatigue specific to the seafaring industry, there is much that can be learned from guidance for other sectors. For example, the UK's Health and Safety Executive has produced various guides on managing fatigue; it identifies the UK Office of Rail Regulation guidance as being transferable to other safety critical industries. This includes the advice below on recognising the signs and symptoms of fatigue.

Likely level of fatigue	Signs/symptoms
Early warning signs of fatigue which should prompt people to look out for more conclusive evidence of fatigue	Fidgeting Rubbing the eyes

Signs of moderate fatigue suggesting performance is being affected. Take these seriously – it is not necessary to fall asleep to make a critical error	Frequent yawning
	Staring blankly
	Frequent blinking
Signs of severe fatigue. Liable to brief uncontrollable micro-sleeps, risk of errors very high	Nodding head
	Difficulty keeping eyes open and focused
	Long blinks

Seafarer fatigue: The Cardiff Research Programme

Fatigue has been extensively researched, particularly in parts of the transport sector such as aviation and road haulage. However, despite an awareness that seafarers might be at particular, and perhaps greater, risk of fatigue than other workers because of the way that they work, until recently there has been very little research focused on them. The Cardiff Programme was one of the first studies intended to begin to fill this gap. It was designed to begin to build up a knowledge base on seafarer fatigue to:

- Predict worst case scenarios for fatigue, health and injury
- Develop best practice recommendations
- Produce advice for seafarers, regulators and policy makers.

To achieve this the programme included a questionnaire survey of working and rest hours, and physical and mental health; a diary survey in which seafarers recorded their day-to-day sleep quality and work patterns; and onboard assessments of alertness and performance (such as reaction time).

The questionnaire survey identified a large number of specific aspects of seafaring that were associated with fatigue. These included, for example, poor sleep quality; negative environmental factors (such as heavy seas and poor weather); high levels of job demands and stress; frequent port visits; exposure to physical hazards (such as fumes and noise) and long working hours. The survey showed that it was those who reported the greatest number of these factors that were most at risk of fatigue. In addition, seafarers who were fatigued were more likely to report having poorer health, poorer well-being and reduced concentration levels and to report having been involved in a collision than seafarers who were not fatigued.

The diary survey showed that fatigue increases most steeply during the first week of a tour of duty. It then steadies but remains relatively higher than it was at the start of the assignment. In order to consider recovery from fatigue, the survey also extended into leave. This showed that fatigue typically does not reduce to pre-tour of duty levels until the second week of leave.

The onboard assessments showed that particular aspects of seafaring work, identified in the questionnaire and diary surveys as being risk factors for fatigue, had a detrimental impact on seafarers' levels of alertness and performance. For example, exposure to noise, working at night and a greater number of days into the tour of duty were all associated with lower alertness levels and poorer performance.

The Cardiff Programme showed that the potential for seafarers to experience fatigue is high because of the number of fatigue risk factors they are exposed to, many of which are unique to seafaring. More significantly, however, it made clear the importance of considering fatigue risk factors in combination – which of course, reflects the reality of seafarers' day-to-day working experience. The consequences can be felt by individual seafarers in the short-term, in terms of fatigue symptoms including, for example, confusion, tension and loss of concentration and in the longer-term as poorer physical and mental health and reduced well-being. Fatigue risk factors also impact on vessels, crews, cargoes and the environment, for example as a result of collisions.

Accident reports and fatigue

Following on from this early research, an initial study in the CASCADe project (Model-based Co-operative and Adaptive Ship based Context Aware Design, 2012-2015) involved a review of official reports to identify the most common factors that lead to accidents. The complex interplay between various factors, including manning and fatigue, is highlighted by the reports.

Accident reports taken from the UK's Marine Accident Investigation Branch (MAIB), the French Marine Accident Investigation Office (BEAmer), the Transportation Safety Board of Canada (TSBC) and marinecasualty.com were studied. In total, 22 grounding and 22 collision reports were reviewed.

Key factors that contributed to the collisions and groundings for those involved in navigating were:
- Fatigue: One or more of the people involved in navigating suffered performance deterioration due to fatigue
- Alcohol/substances: One or more of the people involved in the navigation were under the influence of alcohol or other substances that may affect performance
- Illness: One or more of the people involved in the navigation were unwell
- Uncertainty about responsibilities: Bridge team members were confused about who was in charge of navigation and/or each person's tasks
- Communication onboard: Navigators misunderstood or did not share vital information. Language barriers, cultural differences, personal animosities or a clash of personalities can hamper communication
- Poor communication between different ships and/or ashore: Relevant navigational information was either not shared between vessels or misunderstood, including the inadequate or confusing use of optical or acoustic signals

Chapter 1
Navigation Accidents and their Causes

- Distraction: Members of the bridge team became preoccupied with matters unrelated to navigation
- Poor bridge design: Less than optimal bridge layout or positioning of displays hampered navigation
- Inadequate means of navigation: Essential navigation equipment was dysfunctional or absent, including lack of adequate charts or acoustic alarms and malfunctioning radar
- Inadequate use of navigational aids: Incorrect or insufficient use of navigation and collision avoidance aids, including not using an available ARPA or choosing an inappropriate mode of display on the ECDIS
- Overload: Navigators became cognitively overloaded and lost situational awareness. This differs from distraction as they were exclusively occupied with navigation
- Alarms suppressed, ignored or misinterpreted: Acoustic alarms were in place and functional, but they were either switched off, ignored or misunderstood
- Inadequate manning: The vessel and/or bridge team was understaffed
- External pressures: The crew was preoccupied with (and potentially distracted by) pre-existing problems not related to navigation causing a background level of stress
- Poor weather conditions: Reduced visibility due to fog, rain or other factors contributed to the development of the accident.

Two important conclusions can be drawn from this list. First, accidents are caused by multiple inter-related factors. For example, a lack of personnel may lead to an accident when weather conditions have been poor, as the crew have become fatigued. Fatigue in turn will lead to an increased likelihood of distraction, poor communication and overload. Secondly, in terms of reducing the likelihood of accidents, it's important to distinguish between causal and symptomatic factors. While fatigue is a cause of accidents, it sits in a chain of events and is rarely the initial, triggering factor. By contrast, core systemic factors such as manning need to be the focus of intervention.

Manning and fatigue

The failings that lead to maritime accidents arguably start at the legislative level, where flag states are effectively operating according to market forces and competing for ship registrations. When flag states are competing, it is inevitable that standards will fall as the main incentive is to attract vessels to the flag. A flag is unlikely to attract vessels if their regulatory framework is more stringent than that offered elsewhere. A key component in terms of fatigue is the choice of flag in relation to manning levels. Where administrations have competed for ship registrations, inevitably the number of crew required to safely man vessels has fallen – and in some cases, to dangerous levels.

A ship with insufficient crew may cope during periods of plain sailing, but without any redundancy in the system there is little protection against the extra demands of intensive periods of operation, for example going into and out of port. A crew, without sufficient rest after a fast port turn-around, is extremely vulnerable to the effects of fatigue. Targeting fatigue as the problem, however, misses the bigger picture, and the structural failings that have led to the dangerous situation occurring. Indeed, many of the solutions put forward to reduce fatigue are unlikely to deal with extreme situations like the one-man bridge. In this scenario, the watchkeepers work longer than they should; they often have no backup; and it is a context where identification and management of fatigue is often impossible.

A further challenge is the varying modes of operation onboard a vessel. The number of crew required for open sailing is far fewer than required for port turnaround where there is the increased stress of pilotage, port officials wanting to see the Master, pressure to get away as quickly as possible and the additional workload associated with loading and unloading cargo. If a vessel is crewed for port turnaround then there may be redundancy in the system when the same vessel is sailing in open water. Where such redundancy equates to money, however, it will always be a challenge to persuade shipowners to crew for these intense periods of operation – unless their competitors are required to do the same as well.

One of the problems in building a case for more crew on certain vessels is that current methods for measuring working hours are ineffective. In the Cardiff University Seafarers' Fatigue study 40% of survey respondents reported at least occasionally under-reporting their working hours. If ships present perfectly completed working hours sheets without any legislative violations, there is no cause for the industry to take action or be concerned. We know that working hours sheets cannot always be considered reliable, however, and therefore it is safe to assume that dangerous situations are occurring effectively behind closed doors.

In terms of addressing fatigue, the proper measurement of working hours is essential. If tools could be introduced to objectively identify violations in working hours, then a strong case could be built for increasing crew levels on those types of vessel where accidents are regularly occurring. Only once the full extent of the problem is known, can measures to address it be introduced.

The role of communication

The CASCADe project highlighted bridge communication as an area for change to improve safety and efficiency. Poor communication could indicate fatigue so providing tools to help crew members communicate, particularly in times of stress, could help mitigate the negative impacts. The value of targeting communication is that it is an area that can be improved through the use of technology.

Other factors, such as weather or traffic conditions, are difficult to predict or control, but familiar patterns of communication breakdown emerged in focus groups and surveys conducted on the CASCADe project. These problems included:
- Language-based challenges. Non-native English speakers may misunderstand each other and native English speakers may speak too fast for non-native English speakers to comprehend

Chapter 1
Navigation Accidents and their Causes

- Local language used between a pilot and harbour authorities, so that the Master is unaware of what action is being taken, for example by tugs
- Pilot and Master discussions not shared with the whole of the bridge team.

The way forward

A main recommendation from the Cardiff Programme was that seafarers' fatigue should be treated as a health and safety issue. The causes of fatigue are complicated with many consequences so different solutions are required. Specific, isolated recommendations to one level of industry will have limited effect. So, for example, it might be suggested to crew members that they ask for assistance if they feel fatigued. A broader perspective would recognise that this may not be possible if the ship is under-manned.

The reason the ship is under-manned may be because of market conditions and competition with other companies which operate with fewer crew. The reason for ships having fewer crew may be because of competition between flag states for registration, which has allowed manning levels to decrease. Recommendations which naively ignore these wider factors will be of little value to the industry. At the highest level, international legislation is essential in combating excessive working hours. The evidence suggests that existing efforts have been inadequate so far.

Establishing standards both for measuring fatigue and for recording and auditing actual working hours would undoubtedly accelerate progress. As research has shown, fatigue is much more than working hours, but knowing how long seafarers work for is crucial to evaluating how safe operating standards are. But current working hours recording systems have been shown to be inadequate and it is now essential that management encourage and monitor accurate recording of working hours.

Since 2015 minimum safe manning justification has been placed upon the ship operating company. The company is required to ensure that each ship is manned with qualified, certificated and medically fit seafarers in accordance with national and international requirements. The ship should also be appropriately manned for safe operations onboard. Prevention and management of fatigue is clearly an important aspect of health and safety at sea.

Every individual and every organisation in the shipping industry has a role to play in terms of raising awareness of fatigue. Ship owners and operators and trade unions should provide training for seafarers and empower them to speak out without fear of repercussion when fatigue is, or has been, a serious concern.

It must be recognised, however, that in a competitive market, any one operator has limited freedom in terms of increasing crewing numbers. The only way to impose such a level playing field would be through international legislation. Companies can, however, set standards of best practice which act as a model for others in the industry. Beyond any moral obligation, a case can be made that being proactive in terms of fatigue makes good business sense, especially in light of certain high profile accidents.

Work organisation clearly plays a crucial role when there is a risk of fatigue. This is true both in terms of management of work and activities on board. For example, from a management point of view there needs to be consideration of increased manning in port or during inspections. Onboard measures should take several forms. It may be possible to anticipate fatigue at planning stages and take appropriate preventative action. If fatigue does occur then the crew must be able to call on the Master or other forms of support. Appropriate communication is also necessary in order to identify and rectify the inevitable errors associated with fatigue.

Finally, seafarers have a role to play in terms of both recognising their own fatigue level, and that of others. In particular, a challenge can be made to the culture of not admitting when one is tired. In terms of specific interventions, a cup of coffee or some fresh air can undoubtedly help, but it has to be acknowledged that such low level interventions will have limited impact in the face of some of the wider industry challenges described above.

Conclusions

While a seafarer falling asleep and having an accident can ultimately be attributed to fatigue, it is critical to understand that multiple factors have led to this scenario occurring. Interventions, such as bridge alarms, have their place but may encourage a view that fatigue needs to be managed, rather than being seen as a symptom of many other systemic failings.

Thanks to recent research on manning, fatigue and accidents at sea, we now have knowledge that will help us to prevent and manage fatigue. In order to rectify the situation it is essential that all those involved treat fatigue as a health and safety issue and use a variety of approaches including technological advances and changes in operational practices. Assessment of appropriate manning levels is likely to be an important long-term strategy which can only occur when working hours and fatigue levels are properly audited.

Aztec Maiden (image: Ulf Teske)

Chapter 2

The need and value of passage planning

By Captain Nicholas Cooper

It can be argued that most groundings and collisions are due to poor risk assessment, and therefore good passage planning is the single best guard against those incidents and accidents. Passage planning is taking the time in advance to identify risk and to agree, as a bridge team, the best way to mitigate against them. There are often unexpected situations at sea during voyages that need a quick reaction and the skills, knowledge and experience of the Master and crew. However as the old adage states, forewarned is forearmed and advance preparedness can save many close calls.

SOLAS Chapter V – Regulation 34 requires there to be a passage plan which ensures that all phases of a passage are considered from a risk assessment point of view. Each phase will have its own risks and challenges that need to be considered and managed. These will include open ocean, coastal, river and harbour transits.

Planning failures

Statistics show that most groundings and collisions are due, at least in part, to poor planning and failure to anticipate risks. Common causes of groundings are a failure to accurately predict under-keel clearance (UKC), being caught too close to a lee shore when weather gets bad or dragging anchor due to insufficient alertness to predictable environmental forces. Collisions are often caused by congestion when more sea room could have been arranged during passage planning, or if the risk of distraction was not anticipated when other operational considerations could have been predicted.

The planning process is also an invaluable opportunity to strengthen the effectiveness of a bridge team when pressures of navigation and decision-making are at their highest. Sharing plans, jointly predicting risk, and benefiting from the experiences of bridge team members of all ranks, will help develop situational awareness within the team.

Sharing a well thought out plan with the pilot will also help strengthen that working relationship and improve safety going into and leaving port, as well as berthing and unberthing. A frequent finding of accident investigators for incidents in port is that there was not a common and shared plan between the pilot and bridge team that allowed for effective monitoring. Inadequate sharing of important information is another feature of poor planning.

Chapter 2
Navigation Accidents and their Causes

Identification of risk

As a minimum, the passage plan should identify all navigation hazards adjusted for anticipated water heights and squat. It is not enough to merely calculate UKC along the centre of the channel and along the keel. It is essential to take into consideration the operational width of the channel and the full beam of the vessel (see Chapter 4).

Knowing your squat is probably one of the most vital elements of your passage plan. It should be stated in bold on the very front page of the plan, the pre-arrival checklist and the pilot card. Your draught plus squat minus available depth tells you how close to the bottom you are likely to be at any stage of the voyage, right up to the berth. Squat curves should be displayed at several prominent locations around the bridge; then, if you know your speed, you can calculate your squat. The front page of the passage plan should also have brief details of the company's UKC policy for when the vessel is under way and alongside.

To illustrate the importance of squat, take the case of a Master who made no allowance for squat at all. He was instructed by charterers to arrive at a river port with a 12.0 metre even keel fresh water draught, which he did to within a few centimetres. The pilot, after boarding and before entering the river, had shown the Master a very recent bathymetric chart indicating that there was 12.0m of water in the channel, so the Master agreed to proceed. After he had run aground in the river the subsequent scrutiny revealed that neither the passage plan, the arrival checklist nor the pilot card made mention of squat, which a simple calculation showed to be about 1.5m at 12 kt!

It is also worth anticipating weather windows and anticipated traffic, adjusted for the time of day and/or visibility. Passage plans should take into account and document anticipated methods of positioning (visual, radar, GNSS) and the intervals between fixing. If GNSS is primarily used, for example with ECDIS, it should be documented how this can be checked by radar overlays, parallel indexing or even beam bearings.

Accidents are often the cause of work overload, fatigue or just tiredness, which can often be anticipated in advance and documented as part of a passage plan. It is often possible to predict distractions due to increased reporting demands, traffic at congestion points, channel crossings, fishing areas, or even leisure craft on weekends. Where possible, passage plans should anticipate areas where it is likely that extra lookouts may need to be posted, where an extra officer is needed or where the Master should be called.

Case study

Ever Decent and *Norwegian Dream*, English Channel, August 1999. The *Norwegian Dream* had only one officer on bridge watch although company procedures called for two in areas of heavy traffic. The officer was also distracted by routine tasks during the material times.

The report from the Bahamas Maritime Authority in May 2000 said the Captain of the *Norwegian Dream* "failed to realise that the area in which the accident occurred had the potential to produce a very heavy burden on an OOW". Many ships use a RAG or red, amber green alertness rating, where in red (high risk) situations the Master is called and no distractions are allowed.

It is also possible to anticipate tiredness or fatigue when demanding operational tasks, such as cargo, berthing, audits or inspections, are expected. Known areas of heavy weather or beam seas or swells, can also be fatigue inducing and require extra manning arrangements, if feasible. In some cases it is possible to take on an optional pilot. Not only does the extra expertise on the bridge reduce fatigue and improve safety, it also offers a unique opportunity for sharing information or mentoring. See Chapter 1 for information on fatigue and Chapter 12 for information on mentoring.

Sources of information

Passage planning starts with the management of information. Traditional sources of information include charts, Sailing Directions, company history and, of course, officers' experience. All of these sources are still highly valuable. However, modern information systems add new sources including websites (such as those of ports and pilots), electronic tools from hydrographic agencies, weather routeing services and even real-time logistics planning agencies.

Prudent navigators will use all means available to learn as much as they can about a port if it is the first time the vessel has called there. Local agents and pilots are usually good points of contact for local knowledge. Ask your fellow officers and crew; get them to listen or talk to other vessels at anchor or waiting for a berth and monitor the AIS for vessels already alongside to identify, for example, container, bulk or petro chemical berths where your vessel may be berthing.

Masters must, however, understand that as the sources of information increase, it becomes more important to be aware of the accuracy of the information. Hydrographic offices (HO) usually have robust quality control processes. However, websites and hearsay should be taken with a degree of caution. Increasingly curious sources of information are meteorological organisations, some of which offer free services although others charge for experienced interpretation.

Anything that helps mitigate a risk can go into a passage plan. However, it is essential to ensure the volume of information does not negate its effectiveness and, most important of all, that all members of the bridge team understand the full passage plan, including the use of language and symbols.

Executing and monitoring the plan

Once a voyage is underway it is essential to monitor progress against the plan. This is best done by all the bridge team as undetected errors by one person are often the cause of casualties and accidents. On most merchant ships a single OOW usually makes all decisions for avoiding collisions and conducting safe navigation for about 80% or more of the time. Humans can never be expected to be 100% error free. It is therefore very important to recognise risky situations and take precautions.

Mitigation techniques include making best use of the lookout as part of the team, calling the Master, calling another officer and the effective use of alarms and alerts. Errors will happen and the successful bridge team will capture these errors before they become incidents.

The effective use of alarms and alerts can often save the day. Overuse of uncoordinated alarms, particularly on different non-integrated systems, can be a huge distraction for navigators, an issue recognised by the IMO as a priority for change. However, some of the more useful alerts that can be used are the safe water contour and UKC in ECDIS, the CPA alert in radar and waypoint alerts.

> **Case study**
>
> *Costa Concordia*, Italy, January 2012. Grounded on rocks and 32 people died. The most complex and costly salvage operation ever ensued. (See Chapters 3 and 4, pages 19 and 35)

In his assessment of the *Costa Concordia* incident, in *Bridge Resource Management: From the Costa Concordia to Navigation in the Digital Age*, Antonio Di Lieto points to issues that he says contributed to the grounding. This includes inadequate training for the ship's integrated navigation system, ambiguity from using electronic and paper charts and confusion due to the use of two languages on the bridge.

More generally, inappropriate scales or zoom levels on ECDIS, which can result in navigation hazards not being clear, have also been regularly noted by investigators as a contributing cause to incidents.

It is essential that all officers feel confident to work with the pilot, feel a shared sense of ownership of the passage plan in port and monitor the plan effectively. ECDIS can be a very effective tool for monitoring passage plans in port given its real-time positioning. The OOW should know the intended route, upcoming waypoints and hazards and be able to question the intentions of the navigator or pilot in ample time to avoid an incident. Should the OWW not be aware of the intentions, the ability to monitor and intervene is lost and is often cited as a leading cause of accidents.

Common faults in planning

Passage planning should not be considered a tick box exercise using the ISM or previous plans as templates. This is a widespread fault and can lead to complacency.

It is not uncommon to find, following a grounding, that waypoints have not been properly checked and when plotted as recorded in the passage plan, they are on dry land! Waypoints are the signposts for the voyage which should have been selected on the basis of the safest route allowing for all the precautionary measures discussed in this and other chapters. Do not use waypoints from previous plans as navigation features may have changed. It has often been noted that when the course and boundaries of channels have been changed and buoy positions altered, ships still follow the original route, reusing waypoints from earlier plans.

Reference to the required nautical publications must mean *relevant* publications and not outdated ones that are no longer valid for the current voyage. It is no good referring to Sailing Directions which have been superseded or charts which have not been corrected up to date. It is necessary to ensure that all onboard sources of information are up to date. Paper-based products such as charts and books should be clearly marked with updates; it can be more difficult, yet absolutely essential, to confirm that electronic sources such as ECDIS are current and up to date.

Passage plans are essential aids to navigation. They must be taken seriously and prepared diligently for every voyage. Risk evaluation and avoidance can only be addressed with such an attitude.

Some personal experience and recommendations

Let the OOW execute the passage plan

In busy waterways like the Dover and Singapore Straits, I relied on the OOW to keep track of the passage plan details and to call out the distance count-down to the next waypoint. This chapter discusses passage planning, so I will try not to stray too far into the subject of bridge team management but the two are linked. Masters who try to keep track of every detail in the plan, including trying to handle all the VHF traffic to and from different entities, are likely to find themselves victims of information overload and distracted from the task in hand, which is overall management of the passage.

But while we are on the subject, let's go into a bit more detail. You must include your lookout in the bridge team and they should be encouraged to move around the bridge, and have the confidence to tell you not only what they are seeing, but what they think it is and what it is doing. Show them how to read an ARPA plot and explain the AIS and ECDIS so they can understand them. If you engage them from the start they can be an invaluable backup to the Master and OOW.

Unless things get really intense, perhaps in the Dover or Singapore Straits, let the OOW run the whole show including navigation, traffic management, radar, ECDIS, VHF calls and reporting. This leaves Masters free to monitor the whole situation and guide and advise as necessary. Masters can then pick up on any mistakes made without taking the task away from the OOW. This is the way the major cruise line operators are training their bridge teams, and is something I successfully adopted for over a decade. As pointed out in many investigations, undetected errors by one person have been the cause of many casualties and accidents. Words of encouragement and well placed compliments go a long way to boost confidence in junior officers who have never been given this level of control of the vessel before.

If, as Master, you insist on running the whole show yourself without full engagement with the bridge team, you run two major risks. They might learn your mistakes or errors as good practice. Even if they do spot mistakes they might not have the confidence to let you know about them. In either case, you will be failing the very purpose of the passage plan, which is to take the vessel as safely as possible from port to port.

Minimise risk

We, as Masters and OOW, are not in the business of taking risks. We are in the business of avoiding, minimising or mitigating them. It might not have occurred to you, but your passage plan is one long, joined-up risk assessment. At each stage of the plan you will have assessed the risk and will have taken one of several corrective actions.

You could avoid a risk by, perhaps, going round the outside of an island instead of between it and the mainland, thus avoiding traffic choke points and inshore traffic.

You could minimise risk by passing a couple of miles further off that cape or headland. A classic example is longitude 18° 00´ West off Dakar, which every navigator in the world, except us, chose when running up and down the West African coast. There was a constant barrage of "ship on my port bow/ship on my starboard bow" on the VHF, and we could see the numerous near-misses clearly on radar and ECDIS as we cruised serenely up and down at 18° 20´ West, unaffected by constant end-on situations day and night.

Some high risk areas like the Singapore Strait cannot be avoided, but you can mitigate the risks. Mitigation measures could, and should in my opinion, include coming off the main engine load programme to manoeuvring full ahead. Anyone who proceeds through there at full sea speed without being able to reduce speed at a moment's notice is asking for trouble.

Double the navigation watch

As part of the passage plan you should double the navigation watch and lookouts, two radars (of course) on 6 and 3 miles off-set, although watch out for vessels creeping up behind you. One OOW would concentrate on navigation with five minute plotting intervals using one of the two radars for this purpose. They will also keep track of the plan and keep a rough log. The other OOW will handle communications with VTS and other traffic, and otherwise concentrate on traffic management with me. I was not one for fixed or rigid roles, and I encouraged a degree of flexibility when any one of us could fill the gap for the others if needed.

The whole plan might fail though if you don't communicate with each other, which means, for instance, that the navigation OOW won't alter course without telling me, and I won't make collision avoidance alterations without telling them.

Chapter 2
The need and value of passage planning

Owners' role

Prior to starting the voyage, the passage plan should be shared and understood by the entire bridge team and where appropriate other stakeholders such as the engineering staff. Most companies, as part of their SMS, now require the passage plan to be submitted to shore management for the purpose of quality management and in the unlikely event of an incident. Be clear that the ship owns the passage plan. It is not for the shore management to approve the plan or speculate on the details. If shore managers have questions about the plan they should discuss it with the Master.

Conversely, the plan does not create a rigid contract for the ship. Masters and bridge team members must have the ability, under the authority set out for them in the ISM Code, to deviate from the plan using their professional judgement and risk assessment. Far too many cases have been reported where ships' crews have felt unable to accept advice from pilots or VTS operators as they preferred to follow a shore approved berth to berth passage plan. This is wrong and is highly likely to lead to collisions or groundings. It has also become common for inspectors to find fault or issue non-conformities to vessels for deviating from a passage plan based upon pilot advice. This is also wrong.

The future

The future is predicted to get busier, with more and larger ships navigating with increasing numbers of non-ship water users including those servicing offshore energy, aquaculture and mining; autonomous or drone vessels; and leisure users. All of these will create risks which need to be anticipated and mitigated, ideally through passage planning and, hopefully, intelligent marine spatial planning (MSP).

The amount of available data and information sources will also increase. Guarding against information overload will become more of an issue for navigators who will need to develop better information management skills. The goal should be to have clear and relevant information presented when you need it, for the task at hand. With the increase in sources of information the quality and reliability of information will also become more important, particularly in relation to the measurement and calculation of safe water.

It seems that by default, greater reliance is being put on GPS and other GNSS (Glonass, Galileo and Beidou) all of which share common faults. These systems are prone to deliberate or unintentional jamming and spoofing. A good passage plan should identify how to detect a GNSS fault and how to cope with it. GNSS loss drills onboard are an excellent idea. Be prepared for alarm chaos and fade of traditional skills.

There are services being developed to provide ships with ready-made passage plans from shore organisations. The benefits include information that may be more up to date and better managed. However, it is still essential for the whole bridge team to be familiar with the plan as ultimately risks will be managed by the Master and the team onboard. There needs to be adequate time for all navigating officers to understand and own the plan. This is a challenge that needs to be met and overcome.

Chapter 2
Navigation Accidents and their Causes

There is also likely to be greater monitoring and management of traffic in congested areas from shore operators. Advancements in IT and communications technology may make intervention from shore authorities possible. However, regulations and the law change more slowly and for the foreseeable future Masters and their crews will maintain the authority and liability for the safe navigation of the vessel.

We are also in a period of unprecedented advancement in the use of technology. As new systems and services become available many ships will be using the dual systems of traditional and modern approaches. We should retain the best of the old and adopt the best of the new. However there are many risks with the use of dual systems or techniques which can cause confusion and a greater workload for bridge teams.

Conclusion

Passage planning is probably the best risk assessment practice for avoiding collisions and groundings. Poor passage planning and single person errors are often cited as a major cause of accidents.

It is essential that bridge teams have a common understanding of the plan, how it should be monitored and when and how to deviate from the plan *when* it is required.

The basics of good passage planning can and should be taught in STCW compliant courses. However, good passage planning is a skill that needs to be practised and can always be improved. Passage planning and monitoring should be a team exercise and will present a range of essential mentoring opportunities for all. Masters have much to pass down, and juniors may have ICT skills that can be passed up.

Chapter 3
Bridge resource management

By Captain Robert Hone

Good bridge resource management (BRM) is a culture that needs to be embraced, both at sea and ashore to ensure the safe execution and completion of a vessel's voyage. Good BRM ensures navigators make the best possible decisions using all available information, resources and assistance. If mistakes are made they should be caught in time and lead to nothing more than a lesson learned, rather than a major catastrophe.

I have imagined a series of events to show how important good BRM is in avoiding a major incident. At any point in my example you could stop and reflect on what might happen next. Some elements will undoubtedly sound familiar.

It's 0500 and just getting light; the 22 kt, 20,000 teu container ship is due at the pilot station at 0630, the third mate, who is OOW, has given an hour's notice to the engine room and called the Master. It's the crew's first call at Zang Xon and all's well.

At 0530 the OOW plots a GPS position on the paper chart and glances at the radar. The fix looks good, so he adjusts the autopilot a degree or two to bring the ship back onto the red line. There are a couple of small fishing boats around, but nothing to worry about, so he calls the Chinese pilots on Channel 14, but gets no reply.

The Master comes to the bridge just before 0600 and sends the lookout to make him a cup of coffee. At that time the OOW pulls the engine controls back to manoeuvring full ahead before calling the pilot station again, this time with success. The request is for the pilot ladder starboard side 1m above the water. The Master asks if everything is OK, and adjusts the course a few degrees away from the red line explaining that he wishes to make a broader turn into the fairway approach. The OOW relaxes as the Old Man has altered course and taken the con.

As the pilot boat approaches, the OOW takes a VHF and goes to the pilot boarding station, leaving the Master alone on the bridge. On the way down he passes the duty lookout going up with the Master's coffee.

On the bridge the Master takes his coffee and tells the lookout to standby to take the helm into hand steering. A Chinese voice on the VHF tells the ship to slow down because the ship is going too fast for the pilot boat. The Master pulls the engine controls back to slow ahead, and calls down to the OOW at the pilot station for a report on the approach of the pilot boat. He then dashes to the bridge wing and looks over the side. The ship is still too fast. Where is the chief mate?

Chapter 3
Navigation Accidents and their Causes

Looking up he notices quite a few more fishing boats leaving the port, and the visibility starting to decrease with the sunrise – two miles, maybe less. A quick look at the radar shows that he is being set in towards the harbour entrance and is only 3 miles from the breakwater so he decides to stop the engines. Where are the leading marks on the chart? Not there, but no worries, the pilot will be here soon.

As the chief mate comes onto the bridge the OOW advises the Master over the VHF that the pilot is on the ladder, but that the lee is not good and the speed is still too high. "Port 20," orders the Master; the helmsman takes the wheel and steers port 20°.

After a short while the OOW, the pilot and second pilot arrive on the bridge. The ship starts to develop a list to starboard due to its speed and rate of turn. The OOW disappears aft to his mooring stations.

The Master has ordered starboard 20° helm and the ship is starting to roll back across to port. The pilot is speaking into the VHF and the second lookout has just arrived, to be told to get the pilot a drink. "Do you want tea or coffee, pilot?"

The pilot stops talking on the VHF and shouts: "Midships". The Master checks the log speed – the ship is now slowing down but still visibly drifting towards the entrance. The chief mate asks the pilot to confirm where the berth is. But the pilot has returned to the VHF, and so the chief mate asks the second pilot. They go to the chart and the second pilot starts to explain how much Zang Xon has developed recently and that the berth, not represented on this chart, is 25 minutes up river from the port entrance.

The second lookout decides to make some tea and leaves the bridge. The chief mate then phones the engine room to check the ballast tank soundings; seeing the Master busy and stressed, he leaves the bridge. The Master is now alone to manage pilots, shipping and the ship's own approach into the busy harbour.

The Master is not happy. He wants to see the ship's position on the chart, where he is in relation to the entrance and where he is going once inside the breakwaters. This is not a good way to start the day! This is his recurring nightmare – being left alone on the bridge with an unintelligible pilot, no back up, no up to date charts, a ballasting operation underway and nobody to simply answer the phone or cancel an alarm. A stressful, nerve-racking, unprofessional and ultimately potentially dangerous situation has developed.

However, it doesn't have to be like this. Very simple BRM techniques, proper, basic passage planning (see Chapter 2) and the use of good checklists would ensure that the situation was controlled in a proper seaman-like manner, without stress or anxiety. So how can this happen?

We should all know the answer; it's in the Cs – the seven Cs
- Clarity of purpose
- Consistency
- Concise communication
- Conduct and proper use of checklists

- Competence and confidence
- Common ownership of the plan
- Credit and congratulation.

Failure of any of these points can be attributed to most navigational incidents and a breakdown in effective implementation of BRM. Let's rewind for a moment.

Clarity of purpose

All the ship's crew is one team and so must understand the rules and regulations which control its working. These rules may be prescribed by the flag state, will be covered by the ISM Code and the vessel's SMS. Everybody must have an understanding of what they entail. All crew should understand the safety culture which is prevalent at sea. Finally, the crew members are part of the ship and this gives them a common identity. It is their home and represents status, security and purpose for them while onboard. Masters are the team leaders and they should tell crews exactly how duties should be performed and all should know their place in the team onboard.

When I was sailing on a small passenger ship we would hold daily coffee meetings or Coff-Cons as we called them, at department head level. For 20 minutes or so, we would discuss the ship's business and feedback from both passengers and crew. Each of the departments had their own meetings and information would filter up from these meetings and then down to the appropriate crew. The entire crew complement understood that there was a conduit to the executive team, and we knew that our decisions would be cascaded down to the relevant crew members. We were all part of the team.

Consistency

We all have to carry out our jobs in the same way and this guidance comes in the form of the company standing orders and SMS. It is vital that all the members of the bridge team understand exactly how their jobs are to be carried out and the resources available to them. This includes the helmsmen and lookouts. The team should understand and own the passage plan and it is essential that it is adhered to. Any amendments must be discussed and the reason for the change recognised and evaluated. Situational awareness should be inbred into the minds of the bridge team. The report into the *Costa Concordia* tragedy comments on inconsistencies in the operation of the bridge team and there are important lessons to be learned from that.

Concise communication

As crews are increasingly likely to be made up of different nationalities it is essential to ensure that all crew members are aware of the working language. Officers must not give orders in different languages. This can be a particular problem on the bridge

Chapter 3
Navigation Accidents and their Causes

when only two nationalities are present. I worked on a vessel when, even though English was the working language onboard, the officers used their mother tongue for most of the orders on the bridge. This point cannot be over stressed; the only way to learn a language is to use it.

I have sailed on a ship with over 40 different nationalities represented in crew members and English was spoken by all the crew even when off duty. Any crew member would be censured by one of their peers for reverting to their mother tongue if in a public area; in an emergency everybody needs to be able to understand and communicate using a common language.

Conduct and proper use of checklists

In any incident the prompt use of a checklist to give structure and priority on what needs to be done to control the situation is essential. A checklist will provide a series of questions which Masters or their deputies need to address and should prompt exchanges. The key to communicating in an emergency is to keep it simple and truthful.

Clear, competent command and control is essential for the safe resolution of any emergency. The team leader should be aware of all that is happening and not just navigation aspects. Delegation and integration are essential as are the abilities to form a plan, communicate this to the team and monitor its development using checklists as signposts to provide situational control. Again, there are lessons to be learned from the *Costa Concordia*.

Competence and confidence

As team leader, Masters should be above reproach. They are in charge and will be closely watched by the crew onboard. Any indication that crew members' conduct is not as required should be quickly noted. The conduct of Masters reflects the ethos of the crew and they have the opportunity, duty and responsibility to mould the crew – the team – to their requirements.

It is important that Masters have the technical competence and confidence to operate, adjust and understand the operation of the equipment that the bridge team is using. Bridge team leaders must have the competence and confidence to solve problems when they arise to their entire satisfaction. It is here that the resource element in BRM is crucial.

Each individual's input into the team is of equal importance and this should be acknowledged by all. This team ethos will be developed by regular briefings and de-briefings before and after manoeuvring operations. These team meetings need not take too long, but do have to be carried out consistently and with a checklist to ensure that all the relevant points are covered. At this briefing Masters must explain how they want arrival or departure operations to take place. Now is the time to inform the team of the berthing procedures, pilot arrival times and stations. The weather report should be read out, or on smaller vessels discussed between the Master and the OOW, and any last minute information disseminated to the entire team.

Everybody should be given the chance to input into the discussion – who was the last person to have been at this port, and when? Any cultural restraints should be managed in a positive manner, and questions should be encouraged. Of course, the extent of the briefings will differ from ship to ship – that on a small coaster with a small bridge team would be smaller than on a passenger liner, as we saw in terms of the weather report above, for example. However, the guiding principles will be the same. All involved should know the plans, what their roles are and how they integrate into the team.

Case study

Maersk Kendal, Singapore, September 2009. The container ship was southbound crossing from VTIS sector 9 to sector 8 off Singapore. At 0645 an hour's notice had been given to the engine room and the Master was on the bridge. It was going a little faster than they had planned, but the Indian chief officer assumed the Master knew what he was doing. Just after 0700 VTIS called up the *Maersk Kendal* to require it to slow down as three ships were coming out of the Jong channel. A few moments later the VTIS called again and the Master told VTIS he was slowing down, that he knew of the two ships ahead and planned to pass astern of both of them. Three minutes later the *Maersk Kendal* was aground. How could this happen?

The UK MAIB investigation concluded that the cause of the accident was a combination of complacency and lack of situational awareness on the Master's behalf. This was coupled with the Indian chief officer's cultural reluctance to challenge the Master which meant that his role was reduced to passive bystander. There had been a complete breakdown in bridge team and resource management.

Common ownership of the plan

Team briefings and, as important, de-briefings are the key actions for successful BRM. A voyage overview should be carried out with the heads of each department present and a voyage overview checklist should be used to ensure a thorough briefing. This is especially important on smaller ships, when the entire crew may be present for the briefing. Everybody onboard should understand the Master's requirements for the voyage ahead and how the planned voyage will proceed.

On ships with dual watchkeeping (six on six off), the possibility of regular breaks can be discussed. A 30 minute break can help to reduce fatigue, especially if it has been agreed beforehand. OOWs should ensure they have everything required for their watch before arriving on the bridge and at no time should they leave the bridge unmanned for any reason. In my experience on coasters, the bridge can become the central hub onboard with off duty crew coming up to have a coffee with the OOW. This has to be managed and Masters must ensure that OOW are not distracted from their primary duty – the safety of the ship and crew.

Chapter 3
Navigation Accidents and their Causes

According to the MAIB report into the grounding of *K-Wave* at Malaga, Spain, the officers on the bridge were celebrating the third officer's birthday in the form of an impromptu party. The *K-Wave* ended the night aground on a Spanish beach.

Credit and congratulation

On the completion of mooring operations, when the ship is securely alongside, debriefings should take place. These don't have to be prolonged events, but all officers involved in the operation must attend. Comparison should be made between the planned and the actual manoeuvre. Each officer should be allowed to talk through the event, starting with the most junior and finishing with the Master. If it is possible the pilot should also be encouraged to join in.

I have found this the most rewarding team meeting, when an exchange of ideas on the clarity of orders, movement of the vessel and individual experiences can be discussed. This gives Masters the opportunity to get to know their officers better and mentor them in a positive manner.

As we all know, manoeuvring a vessel alongside is not a simple operation and things will not go according to plan. These events should be examined in some detail and the reasons for the changes to the plan understood and acknowledged. Notes should be made and a record made of any physical differences or obstructions in the port or the approaches that are not charted and the appropriate authorities should be informed. This will include advising sisterships that may be using the port in the future.

It is important that near misses and smaller incidents are reported to the ship's owners via the safety officer. All incidents should be considered learning opportunities and reported to the appropriate reporting system. This includes MARS, The Nautical Institute's Mariner's Alerting and Reporting Scheme; CHIRP, the UK confidential reporting programme for the aviation and maritime industries and national maritime administrations. The obvious needs to be questioned; I remember once asking if the stabilisers should still be extended when approaching a berth – they were very quickly retracted.

Any outstanding input from officers should be congratulated. It is most important that credit and thanks for a job well done is expressed at this time. It cannot be overstated that team management needs influential leadership and that a few words of praise and a "well done" are key in developing a strong team ethic onboard. It is of vital importance that the team is a happy one and that the individual's well-being and health (mental and physical) are considered and monitored. The old adage that 'there's no I in team' is true onboard ships. It is paramount that individuals realise they are part of the team and that the team's strength lies in each individual looking out for, and enjoying the trust and support of, all the other team members.

Navigational audits and future trends

Managing the team onboard a ship, and especially the bridge team, consists of planning for several different scenarios. We have looked at the at-sea requirements and hopefully you will remember the seven Cs. Failure to do so has led, and will lead, to navigational accidents.

More often the bridge team is involved with tasks that are outside the normal seagoing experience. These tasks range from dealing with emergencies, cargo operations, drydocking and onboard audits. Potentially there could be unplanned consequences during ship audits and inspections.

Most important is clarity of purpose; all the team members must understand what the Master wants as the final outcome. This clarity comes from focused regular meetings and briefings with all team members. Making a plan, or delegating the plan making to your officers, reviewing the plan and acting on the plan as a team will lead to positive and controllable outcomes. Situation awareness and sound knowledge and experience of all resources available are essential ingredients of any bridge team.

It is important that all team meetings are recorded in the deck log book as evidence of BRM best practice. This record can be noted in the Master's hand-over notes and voyage report back to the shipping company. Masters are team leaders and it is their responsibility to mould and develop the team as they want it to be.

So how will BRM evolve? During my seagoing career I have seen the practice onboard change from a top down, autocratic, almost tyrannical chain of command with nobody daring to challenge their superior officers. Now with BRM it is a dynamic team organisation where every action or order is checked and the safety and purpose of actions confirmed and understood by the whole team.

Some of the more radical ideas concerning BRM suggest a rank-free team with each member having a specific role to undertake, irrespective of their rank on board. This function-based bridge management allows Masters to assign roles to the bridge team for navigators, co-navigators, helmsmen, administrators and operations directors. Carnival UK and Princess Cruise line have been foremost in advocating this approach with some success. The incorporation of the pilot into the bridge team with the function-based method is simple.

The passage plan is berth to berth and so pilot exchanges and briefings take place while navigators continue to con the ship. Only when the operations directors are satisfied will pilots be allowed to input and they may or may not take the con. VTS exchanges and other messages are all taken by administrators, who pass on relevant information. eNavigation and prescribed route planning are discussed at the passage planning stage, but also need to be managed within the BRM structure.

On smaller ships the changes seem to happen more often. Watchkeeping routines can be altered by Masters and in some cases it may be justified to delay the vessel to ensure that the watchkeepers are properly rested. A positive professional attitude and good team dynamic can be developed in junior officers by giving them opportunities to berth ships and by mentoring them. On coasters the team will probably consist of the entire ship's company, so an understanding of equal work load sharing is essential.

Summary

Industry changes signify a cultural shift to BRM, and this, coupled with the increase of automation on the bridge, means that Masters cannot afford to stand still and manage ships in the way it was done in the past. Bridge cams, video and audio recording and live feeds from the bridge back to offices all mean that Masters are accountable for the style of team management they adopt. Their clarity of purpose, consistency and confidence will be reflected in the performance and conduct of the bridge team.

A clear and common understanding and ownership of the passage plan are the building blocks for a successful BRM. This requires a positive, learning attitude from all deck officers. Concise communications can alleviate language problems with officers and crews of different nationalities and feed-back loops should always be used when giving orders.

A good team is a happy team. Masters should actively encouraging professional development in their officers and provide recognition and credit where it is due. They should endeavour, by their own example, to be the inspiration for the next generation of navigators.

Chapter 4

Positioning

By Captain Paul Whyte

Navigators over the millennia have grappled with the challenge of knowing where they are in the world. In truth, the key is not just knowing where you are, but knowing where you should not be. Failure to keep ships safe from collision or grounding could be catastrophic, potentially leading to loss of life and pollution of the environment.

Situational awareness

In the era of the electronic revolution, navigation has veered from art to science. However, the task of navigation remains a vital component of situational awareness, which means being able to identify, process and comprehend what is happening around you, where you are and where you are going. In effect, you either have situational awareness or you do not. It is not something lost if you never had it in the first place!

Case study

Ficus, Bahamas, February 2008. The grounding was due to human error and a lack of situational awareness. The OOW was distracted by pre-arrival preparations and lacked bridge team support. Eventually, the OOW committed the ship to a large alteration of course prompted by the track pilot alarm, without first confirming the ship's position. The OOW also failed to properly monitor the turn in strong winds, and as a result became disorientated; all of which led to the ship grounding.

A good navigator should always know with a high degree of certainty the position of the ship with respect to the track and sea room available, and should never rely on one positioning system. The most appropriate method to use will be based on the accuracy and reliability of each system, in order to keep the ship safe from navigational hazards and prevent grounding.

Familiarity with the bridge equipment and regular cross-referencing between navigation systems not only increases confidence in knowing where you are, but provides stimulation, and early warning of any defect or conflict between positioning systems to break the potential error chain. This rigour promotes alertness and prevents complacency, which in turn enhances situational awareness and professional competence.

Chapter 4
Navigation Accidents and their Causes

Heading systems

Independent of any external power source, the magnetic compass is peerless among navigation instruments. The accuracy of the magnetic compass should be of the order of 1-2°. This relies on a good compass swing to eliminate the magnetic deviation caused by the ferrous ship, which is achieved using soft iron correcting spheres and flinders bar, vertical magnets, and magnets in the fore and aft. Regular cross-checking with other heading reference systems will quickly reveal any problems, such as a wandering or failed gyro.

Today's north-seeking gyroscopic and satellite compasses have been game changers. Gone are the large, unreliable gyro bodies, replaced by compact and reliable miniaturised systems that provide good accuracy, especially with a GNSS input to give automatic latitude and speed corrections.

Positioning systems

How times have changed since the late 20th century, when the accuracy of celestial navigation was between 1-3 nm. Today, we rely on GNSS, which delivers an accuracy measured in nanoseconds, giving real-time position accuracy of 10-20m, which is about one hundredth of a nautical mile. Differential position correction signals can improve GNSS accuracy to less than 2m: an amazing one thousandth of a nautical mile.

Despite the advances in technology, there is little point in relying on high-precision systems if the navigator has not taken account of the age and accuracy of the chart survey and the limitations of the chart scale. This mismatch will be compounded by ill-judged waypoints that reduce the margin of safety and the width of sea room.

The navigator must also allow for the limitations of the different position fixing systems, the risk of equipment failure and the interruption to GNSS caused by solar activity, spoofing, jamming and radio interference.

Global navigational satellite systems

GNSS positioning is fundamental to all modern navigation and communication systems, and is primarily provided by GPS and GLONASS. Other satellite systems under development (in 2015) are Galileo, BeiDou, IRNSS and QZSS. While these systems offer greater flexibility, they also increase dependence on satellite technology and can promote complacency, with over-reliance on a single system. So the modern navigator must take steps to defend against this vulnerability and remain alert to intentional or non-intentional jamming or spoofing.

The conscientious navigator should recognise that the precision of GNSS latitude and longitude displayed to four decimal places of a minute is not representative of the accuracy of the system (Figure 4.1). Of course, having an accurate, continuous position reference system has revolutionised the safety of navigation. But again, remember the challenge is not just knowing where you are, but knowing where you should not be.

THE NAUTICAL INSTITUTE

Chapter 4 | 27
Positioning

Datum WGS-84			
S	10°	32.2621	
E	142°	9.5458	
COG	232°	SOG	12.0 kt

Figure 4.1 A typical GPS display

Systems that are dependent on GNSS are automated radar plotting aid (ARPA), AIS, GMDSS, LRIT, dynamic positioning (DP), ECDIS and the voyage data recorder (VDR).

To improve situational awareness, ARPA, ECDIS and most ECS are able to display a true vector based on the ship's ground course and speed. Better still, many ECDIS are able to display the ship's predicted position ahead as a 'ghost ship', and an 'anti-grounding cone' or 'safety frame'. These improve situational awareness when monitoring turns and the available safe water, particularly in confined waters.

To negate any loss in over-reliance in GNSS, ECDIS allows plotting of bearings and ranges of user-defined fixing points, which is especially critical if experiencing GNSS denial when the system defaults to dead reckoning (DR) mode.

It is important to recognise that ECDIS is simply a chart, albeit an electronic one, with more active features than its paper counterpart. Moreover, ECDIS will continue to function as a chart without a GNSS input, so navigators must regularly practise manually fixing in DR mode in order to gain confidence in maintaining an accurate position without GNSS.

Dead reckoning

DR is one of the oldest forms of navigation. It is the process of approximating the ship's position solely based on the course steered and the distance steamed since the last known position. Historically a well maintained DR was a vital tool used by all navigators to anticipate the ship's position for celestial navigation or making landfall, and was used in conjunction with other methods of plotting a fix, such as the obsolete radio direction finding (DF) or when crossing depth contours.

To improve situational awareness, the modern navigator predicts ahead the most probably position (MPP) matched to a DR, or better still, to an estimated position (EP).

Estimated position

By its nature, a DR position is rudimentary, so an EP refines that prediction further by applying set and drift. To remove ambiguity caused by the misidentification of navigation marks, a plotted fix should always be associated with a DR/EP to improve confidence in the plotted fix, and to refine the next DR/EP (see Figure 4.2).

Chapter 4
Navigation Accidents and their Causes

Terrestrial fixing

Ideally, two lines of position (LOP) should intersect at 90° to minimise the pool of errors. However, to avoid a plotting error or ambiguity of the two fixing points, a third bearing (called a confidence check) should be taken, ideally with a bearing separation of at least 30° to 60°. The closer the object, the less sensitive the bearing is to any observation and gyro error; conversely the size of the pool of errors will be exaggerated by an increased distance or high rate of bearing change.

The fixing procedure requires positive identification of each object derived from the DR or EP and by the use of visual and radar shoot-up bearings and ranges; pre-calculating the fixing points; then recording the observation time and bearing, and finally plotting the fix. The navigator then re-calculates the course and speed to maintain or regain the track, and generate a new DR or EP. Speed is of the essence, as the plotted fix quickly becomes historical, falling behind in the ship's wake, and the proactive navigator must generate a DR or EP to anticipate the vessel's position in relation to any looming navigational hazard.

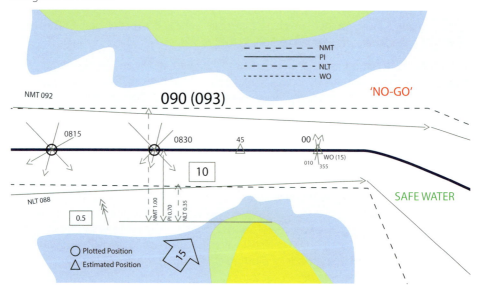

Figure 4.2 Terrestrial fixing, DR and EP

The navigator will eliminate any misidentification, manipulation of the bearings (and/or ranges) to make the lines of position cross, and problems such as the vessel being on the same circumference as the fixing points, provided that:
- The visual fix is not 'blind' by being associated with a pre-plotted DR or EP (actually any fix using visual and/or radar objects)
- Any system errors such as repeater alignment or gyro error are known
- At least three well-judged objects have been positively identified with a good angle of cut.

Radar fixing

The use of radar evolved in the mid-20th century for maritime use for anti-collision and positioning, and can be extremely accurate, providing vital situational awareness in restricted visibility (see Chapter 5). Additionally, in restricted visibility and congested waters, RACONs provide positive identification of navigational marks for radar ranging and parallel indexing.

Digitisation made possible the development of ARPA by processing radar returns to enable computer target tracking. However, this positive advance has unfortunately compromised the fidelity of the radar image for fixing, and the navigator still relies on positive identification of conspicuous radar targets, which are not vulnerable to interference from heavy seas and weather, terrain masking or changes in sea level caused by the height of tide.

As a reminder, the current IMO performance standard requires a radar bearing accuracy of +/- 1° and a range accuracy of 1% of the radar range scale in use or 70m, whichever is the greater. Additionally, correct picture and heading alignment is vital (see Figure 4.3), and the radar beam width and pulse length noticeably influence accurate radar bearings and parallel indexing techniques. So essentially, radar is very accurate in ranging and should be the preferred line of position, and radar bearings should be considered as a last resort.

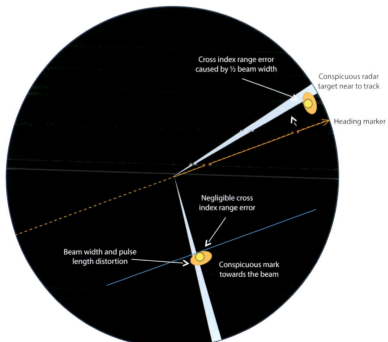

Figure 4.3 Radar beam width and pulse length distortion

Chapter 4
Navigation Accidents and their Causes

Parallel indexing

Parallel indexing (PI) is a system of radar track control that overcomes constant reference to the paper chart (see Figure 4.4). A common error with PI is the misidentification of the radar-conspicuous object – this must be positively identified by bearing and range from the present estimated position.

The selection of a good radar-conspicuous object must not be overlooked. A small cross-index range will be inaccurate due to the half-beam width distortion (see Figure 4.3) and is vulnerable to bearing error caused by any heading marker or display misalignment, or if any gyro error is present. This weakness eliminates the use of breakwater entrances, channel markers or headlands close to the track, leaving the only useful radar-conspicuous objects broad on the bow and towards the beam.

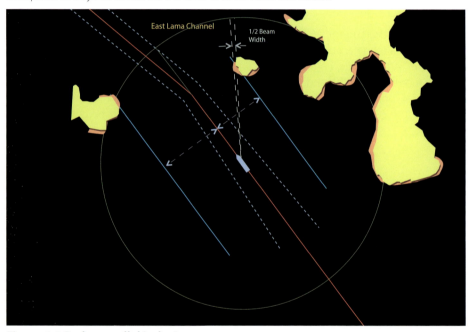

Figure 4.4 Radar parallel indexing

Case study

Atlantic Blue, Torres Strait, February 2009. Grounded during a night-time transit. An unexpected route change meant that two navigational charts necessary for the transit of the Torres Strait had to be supplied by the pilot, and that delayed completing the passage plan. Wishing to avoid conflict with the pilot, as experienced on a previous transit, the Master ordered that the GPS waypoints and cross-track error should not be plotted, so as to declutter the radar display. This removed a critical method of monitoring the vessel's passage plan. Nonetheless, the

OOW did plot positions at regular intervals using visual bearings and radar ranges that were periodically cross-checked with GPS positions (see Figure 4.5).

Crucially though, no allowance was made for the 25 knot north-westerly wind and the east-going tidal stream, resulting in the ship setting south off the planned track. The OOW and pilot had not discussed or defined the off-track limits. Eventually, the pilot ordered a series of small course alterations that amounted to a curve of pursuit, in that they were insufficient to regain track. Unwisely, the OOW deferred to the pilot and did not call the Master, and eventually, the ship grounded on a sandy shoal on Kirkcaldie Reef.

This grounding exhibited the classic error chain of an organisational accident when a sequence of undetected human and system failures led to a seemingly random accident. Track monitoring failed despite using visual bearings, radar ranges and periodic GPS positions and trusting the pilot's visual and radar monitoring techniques. Poor communication, hindered by a lack of double-checking between position systems, and the OOW deferring to the pilot, resulted in the lost opportunity to break the error chain.

A more complete passage plan using all available tools, including parallel indexing, would have given early warning of danger and improved situational awareness.

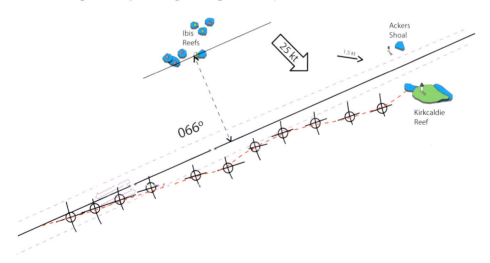

Figure 4.5 Navigational chart used by the *Atlantic Blue*

Celestial navigation

Celestial navigation remains a valid means of positioning, capable of an accuracy of one mile or better when using morning or evening stars, is an excellent back-up to GNSS and for checking the compass error, particularly when offshore. However, poor weather

Chapter 4
Navigation Accidents and their Causes

inhibits observation of celestial bodies and the horizon, and relies on regular practice to maintain competence.

Other systems

The echosounder is an excellent complementary instrument to confirm the ship's position when crossing depth contours, which is a valid LOP that can be transferred to cross with another contour.

Although Loran has been largely decommissioned, its potential successor, eLoran, is currently being explored by a number of countries, including India and the UK, as a valuable back up to GNSS.

Multi service or integrated receivers have been developed to accept input from all types of GNSS, eLoran and other yet to be developed systems. However, integrated GPS/GLONASS systems suffer the same vulnerabilities, which require the navigator to cross-check between positioning systems.

Inertial navigation systems use accelerometers and gyros to measure the movement of the ship from a known reference position, and automatically generate an EP with a growing pool of errors.

Specialist close-in systems such as DP require multiple independent automated navigation reference systems including taut wire, acoustic transponders and differential GPS. These systems are only useful over short ranges and for specialist offshore applications. Nonetheless, DP systems continuously integrate the different position reference systems to improve accuracy.

eNavigation

Although not here yet, eNavigation is a major IMO initiative aimed to harmonise and enhance navigation systems, and is expected to have a significant impact on the future of marine navigation. The International Association of Marine Aids to Navigation Lighthouse Authorities (IALA) defines eNavigation as: *The harmonised collection, integration, exchange, presentation and analysis of maritime information onboard and ashore by electronic means to enhance berth to berth navigation and related services, for safety and security at sea and protection of the marine environment.*

eNavigation needs resilient position, navigation and timing (PNT) from a continuous position reference system – preferably two – which meet maritime requirements for accuracy, integrity, availability and continuity. Possible solutions are:
- Hardened GNSS
- Enhanced aids to navigation (AtoN)
- Synchronised lights
- Radar AtoN (absolute positioning)
- Non-radio systems such as INS

- Complementary EPFS such as eLoran
- Absolute radar positioning
- Modified Tideland Racon
- Modified Furuno Radar on Alert.

Following a planned track

A common experience shared by some mariners using paper charts is an incorrect course to steer. A simple check of the parallel ruler along the track eliminates this all too common 5° or 10° mental aberration. Other examples of mental irregularity would be entering the wrong course into the autopilot or entering the wrong coordinates or waypoint into the GNSS receiver. The simple task of cross-checking between different heading and position systems would break the error chain.

> **Case study**
>
> *Oliva*, Tristan Da Cunha, March 2011. Position monitoring failed as, instead of monitoring the vessel's position in relation to the surrounding hazards on an appropriate navigational chart, the OOW followed without questioning an incorrect GPS track superimposed on the radar screen, and had no warning of the proximity of the danger ahead.
>
> The main cause of the grounding of the *Oliva* was the ship following an incorrectly plotted great circle route that took it directly over Nightingale Island. The next rhumb line waypoint on the great circle route was incorrectly transposed from an inappropriate small scale British Admiralty chart onto a plotting sheet, which was also replicated in the GPS and ARPA.

So even today, the compass and autopilot remain pre-eminent. Routine monitoring of the progress of the ship along the planned track by regular fixing and dead reckoning, as well as cross-checking between navigation systems using an appropriate scale chart, will save the day. It follows that the gyro compass error and magnetic compass deviation should be regularly checked, ideally once per watch, and especially before coastal and pilotage navigation. Such enduring watch routines are testament to a well-run bridge, and should also include radar performance checks, and such things as monitoring the echosounder settings.

Passive navigation

The accuracy of today's satellite-derived position systems has led to the unintended consequence of passive or monitoring navigation, with the navigator becoming an observer. If managed properly, this is a very positive development, as integration of electronic bridge and navigation systems has given the modern OOW the space and time to improve situational awareness to provide spare capacity for decision-making in busy shipping lanes.

Chapter 4
Navigation Accidents and their Causes

> ### Case study
> *Pride of Canterbury,* south-east coast of England, January 2008 (Chapter 9 page 85). Grounded on a charted wreck. The inattention of the OOW, navigating by eye and employing sporadic plotting, resulted in a loss of situational awareness, exacerbated by poor bridge resource management, and improper use of the electronic chart system, which was not configured to show the wreck.

Navigators should be wary of the practice of planning on paper and then just entering the waypoints into the GPS and ARPA in order to simply monitor the ship's position relative to the planned track, or worse still, to navigate on non-approved electronic charts. In this context, the paper chart purports to be the primary means of navigation, when in reality GPS is the sole means of navigation. More diligent navigators will periodically plot a position found by other means and, when appropriate, regularly employ parallel indexing techniques to confirm the ship's position to provide an independent terrestrial-based position reference system, that is, radar pulses bouncing off *terra firma* rather than relying on pseudo-ranging from space.

Active navigation

The introduction of ECDIS has perhaps had the unintended consequence of compelling navigators take a more active role, as the system requires users to, for example:
- Set the safety contour, safety depth, anti-grounding cone, cross-track distances, alarms and warnings
- Load the correct electronic navigational charts
- Adjust the presentation library and map layers
- Cross-check with radar overlays and plotted lines of position
- Interrogate vector objects to determine attributes
- Understand the categories of zones of confidence (replacing the old fashioned source data diagram)
- Correlate the AIS tracks, if fitted.

Paradoxically, it would seem that the former passive monitoring of an electronic chart, with regular position fixing on the paper chart, has become manifestly easier by simply monitoring ECDIS. In reality, an enormous change in mind set is required to recognise the role change from passive to active navigation.

> ### Case study
> *Ovit,* Varne Bank, Dover Strait, September 2013 (see Chapters 9 and 11 pages 86 and 106). *Ovit* was aground for about three hours, there were no injuries and damage to the ship was superficial. There was no pollution. The ship solely used the type approved ECDIS and the MAIB concentrated largely on its installation, operation and aspects relating to training in its use. The main findings of the report included: The passage plan was unsafe as it passed directly over the Varne Bank; the OOW

followed the track on the ECDIS display, but did not realise his ship was aground for about 19 minutes; the ECDIS safety settings were not appropriate to the local conditions and the audible alarm was disabled; although the lights of the cardinal buoys were all working, and were seen by the lookout, they were not reported; the OOW concentrated solely on the intended track on the ECDIS and ignored all other navigation aids in the area.

There was a complete breakdown in the passage planning idioms of proper preparation, planning, execution and monitoring. Poor passage planning allowed the track to pass directly over the Varne Bank – a track that was not properly checked using the ECDIS check-route function or double-checked by the Master. Inappropriate depth and cross-track error settings were used, the scale of the ENCs in use was unsuitable, the audible alarm was inoperative. In addition, the OOW demonstrated poor situational awareness and failed to cross-check the ship's intended track relative to any dangers to navigation that might be encountered on his watch.

Safe water

As previously mentioned, the challenge is not just knowing where you are, rather knowing where you should not be, which means knowing the safe water.

The navigable safe water is a three-dimensional concept. The under-keel clearance (UKC) is the vertical separation from danger (see Chapter 7, page 57), while the horizontal separation from danger, eg limiting danger line (LDL) or no-go line, must allow for the half-beam width of the ship (measured from the centreline) and clearance of the stern or bow when turning away from danger (see Figure 4.6, page 36).

Case studies

Sea Diamond, Santorini, Greece, April 2007. Grounded on an extension of a charted reef within the caldera of the Greek island of Santorini. The age and reliability of the hydrographic survey were factors.

Costa Concordia, Giglio Island, Italy, January 2012 (see Chapters 2 and 3, pages 12 and 19). Grounded on the Scole Rocks off Giglio Island. The scale of the chart and adequate separation from the no-go line were factors.

The main lesson here is that the UKC is not the only factor in determining the amount of safe water.

Chapter 4
Navigation Accidents and their Causes

Before using a chart, navigators should assess the quality of the survey based on the source data diagram, as not all sea areas have been systematically surveyed to modern standards, or surveyed at all. The chart will have been compiled from the best hydrographic data available, which means that not all shoal areas dangerous to navigation have been identified The chart will have been compiled from the best hydrographic data available, which means that not all shoal areas dangerous to navigation have been identified.

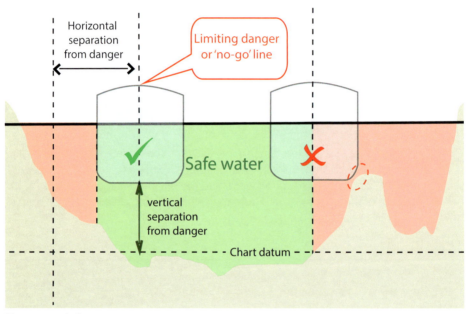

Figure 4.6 Safe water

While electronic navigation charts (ENC) give the illusion of accuracy, they are derived from the same source data as the previous paper version. To assist the navigator to assess the quality of the survey with more confidence, ENCs use a category of zone of confidence encoded against five categories (A1, A2, B, C, D) with a sixth category (U) for data that has not been assessed. The categorisation of hydrographic data is based on three factors of position: accuracy, depth accuracy, and sea floor coverage.

The quality of the survey determines the confidence of the system of plotting a position in relation to the coastline and sea floor, that is, where the ship is relative to danger. Geodesy of the earth plays a role in plotting, as the GPS-derived position must be plotted on a chart corrected to the WGS84 datum, which is especially important if correlating with terrestrial-based methods of fixing.

Finally, the formula used to calculate the UKC is essential for passage planning (see Chapter 2) and will be decided by the accuracy of the survey. In shallow waters, the standard convention is to use a static or dynamic UKC formula that takes account of the height of tide, squat and charted depth, as well as other factors (see Chapter 7). However, if the survey is pre-side scan sonar (earlier than 1973) then a simpler rule of thumb must be used to improve the margin of safety by allowing a clearance equal to one and a half to three times the draught of the vessel.

There are always exceptions to the rule, as experience of well-used straights or channels may facilitate the use of the standard convention despite an old survey. Conversely, estuaries and silting sand banks might require extra caution despite recent surveys carried out with side scan or multi-beam sonar.

Summary

In reality, GNSS failure is a rare event, but over-reliance can lead to complacency in maintaining traditional skills of fixing. Cross-checking between navigation positioning systems might be tedious, but it is vital for navigators to maintain situational awareness and to be ready to manage any primary navigation system failure. Routine is infinitely preferable to an emergency.

So actually nothing has changed over the millennia. Using all available means and diligent routines underpins good watchkeeping. This involves studying the chart and passage plan; cross-checking different position systems (including taking compass bearings and radar ranging); following leading and sectored lights; monitoring clearing bearings and cross-index ranges and monitoring depths. These procedures help develop a clear picture of where you are, where the nearest danger lies, and where you are going – all essential for good situational awareness.

Remember the adage: The challenge is not just knowing where you are, but knowing where you should not be. Modern navigation systems have the benefit of day and night viewing of colour screens, which make possible conferencing and discussion about the decisions to be taken in complex navigational situations. So if there is at all any doubt, call the Master to assist in the decision-cycle.

Take 10: Steps for competent positioning

Be aware and alert: Know where you are – in safe water. Professional navigators need to maintain situational awareness and stimulation to ensure the safety of lives, the vessel and its cargo, to protect the environment and to ensure commercial effectiveness.

Check, check and check again: Good situational awareness requires the continual cross-checking and correlation of positioning systems, so you know where you are and know where you should not be. Use the best scale chart, and know the accuracy of the survey.

Chapter 4
Navigation Accidents and their Causes

Error chain: Never rely on a single means of monitoring the ship's position in order to break the chain of error. Traditional and electronic systems are complementary, giving early warning of system or human failure.

GNSS denial: Although the coordinated use of multiple GNSS, such as GPS and GLONASS, improves reliability, all GNSS share a common weakness and therefore are equally susceptible to intentional or unintentional jamming.

Positioning systems: There are many methods of positioning available to the navigator, some based on traditional techniques and others on modern technology. Using more than one position system is a useful tactic to detect passage planning errors and system failures, and enhances situational awareness.

Human behaviour: Mistakes happen. Positioning methods are complementary, but need to be cross-checked. The professional mariner needs to be the human integrator of these systems, and perishable skills need refreshing and regular practice.

Prepare for the worst: Equipment fails. The loss of GPS is a real risk and can be identified early with routine cross-checking of position systems, supported by contingency plans and procedures for dealing with a loss in any position system.

Avoid over-reliance: Over-reliance on GPS, particularly when integrated into ECDIS, can lead to complacency and poor decisions based on scanty information. The use of GNSS with ECDIS has revolutionised navigation practice, to the extent that it now requires a new approach to active navigation.

Know your stuff: Training in the use of electronic positioning systems is not just about knobs, dials and buttons but, most importantly, how to use technology to support good decisions.

Share your knowledge: Teaching is learning twice over. Embrace bridge team management and encourage mutual respect with those experienced in traditional skills, trading knowledge with technology junkies and explaining protocols and menu structures.

Further reading

IMO MSC.1/Circ.1503 *ECDIS - guidance for good practice*, www.imo.org

Chapter 5

A rough guide to collision avoidance

By John Third

Collisions between ships are an all-too regular occurrence at sea, even nowadays when modern vessels have more sophisticated equipment than ever before. Why should this be?

The simple answer is that the ability to avoid a potentially disastrous close-quarters encounter depends upon the capability of the individual in charge of the ship's navigation. The watchkeeping navigator must have both the knowledge and experience to discharge the responsibility of navigating the vessel. Unfortunately, there are many navigators on the seas who lack the necessary experience and who often apply their knowledge incorrectly.

Experience should tell navigators when a dangerous situation is developing, and their knowledge should dictate how it should be dealt with.

If you hold a certificate of competency as a navigating officer then you should know all about the International Regulations for Preventing Collisions at Sea, usually called the Colregs. However, what you really need to know is how to apply the rules when keeping your watch at sea.

Know your ship

First of all, before taking your place in the watch rota, make sure that you have an understanding of the navigation bridge of your vessel.

Ask yourself:
- Do you know the procedure for switching between autopilot and manual steering systems?
- Do you know whether the maximum steering response is available immediately, or is it necessary to switch on a second steering motor or pump?
- Do you know the procedures for obtaining control of the main propulsion and for slowing or stopping the ship if required?

These basic control measures are essential knowledge for a bridge watchkeeper.

You should also commit to memory particular manoeuvring characteristics as posted in the wheelhouse. You should be aware of the times and distances for advance and transfer in a turn of 90°. In addition to all of this you should know how to operate your electronic navigational aids properly, and you should be well aware of their limitations.

Chapter 5
Navigation Accidents and their Causes

Know your duty as a watchkeeping officer

You are in charge of the vessel and are responsible for safe navigation during your watch. If you have any doubt whatsoever regarding any situation developing, you must call the Master. No Master will criticise an officer for calling them to the bridge unless the call is made too late in a dangerous situation.

Keeping a proper lookout – Rule 5

Keeping a lookout is about being aware of everything surrounding your vessel, including all shipping traffic; navigational hazards such as land, shoals and dangers; weather; wind; current and tidal stream. This requires that you use your eyes to continuously scan the sea area around your vessel.

The function is a process of information gathering and it requires concentration. Every available source must be considered, including what can be seen in the waters around the vessel, what can be heard on the VHF radio and what can be detected using navigational aids such as radar, the automatic radar plotting aid (ARPA) and AIS displays. Keeping a lookout is an obvious and visible activity, but it should also be a subconscious process during which watchkeepers make constant reference to information sources throughout their period of duty.

A proper lookout requires concentration and involves looking ahead, astern and to the sides of the vessel. It also involves moving around the navigation bridge, and you must try to avoid remaining in a static position for too long. Remember you are the guardian of the sea area around your vessel, and surveillance must be maintained over this at all times. You cannot afford to be distracted and you must continually evaluate the effectiveness of what you are doing. If the situation changes and you need to post an additional lookout, then you must do so immediately.

The requirement is to ensure that, using all available means, vessels are detected in sufficient time to enable a proper assessment of their movements and to keep them under surveillance. This approach will enable you to determine whether there is a risk of collision and to decide upon any action which may be necessary.

Remember:
- Do not under any circumstances be deflected from your duty to keep a safe navigational watch
- Do not use a mobile phone or any other portable electronic device while on watch
- Do not let others draw your attention away from keeping a proper lookout.

From these points it should be understood that a watchkeeping officer should under no circumstances become a GMDSS operator. The operation of communication systems for ship's business or other activities cannot and should not be on the watchkeeper's list of duties. GMDSS operation during a watch has been the cause of several collisions, including, in my investigative experience, a very nasty running

Chapter 5
A rough guide to collision avoidance

down of a small cargo ship by a large container ship with the loss of all hands on the smaller vessel.

Management of the lookout function is the responsibility of the watchkeeper and the Master. It is the most important part of bridge operations. Remember if you do not keep a proper lookout, your collision avoidance strategy does not even go beyond Rule 5.

Risk of collision – Rule 7

The rules say that every vessel shall use all available means appropriate to the prevailing circumstances and conditions to determine whether the risk of collision exists. If there is any doubt, consider the risk of collision to exist. In practice, the key to applying this rule is to exercise caution and to assume there is a risk of collision until you are quite satisfied there is none.

A competent watchkeeping officer should know how to determine whether risk of collision exists. The ARPA radar and the AIS are useful aids, but the most reliable 'directly connected' method is to take compass bearings of an approaching vessel. I use the word 'connected' as you should actually go outside and see the approaching vessel, determine what type of vessel it is and build a picture of exactly what you are engaged with. All these factors can be fed into your decision-making process. You should not, under any circumstances, become simply a radar operator looking at 'blobs' and data on the screen. Collision avoidance is much more demanding, and using your eyesight to seek out vessels in the sea area around your vessel is the basic requirement of keeping a proper lookout and navigational watch.

In offshore waters your eyesight may serve you well by detecting small pleasure craft or fishing vessels which your various radar systems have failed to detect. You do not want to suddenly find you are bearing down upon a yacht two wave crests distant and directly on the trajectory path of a course alteration you are making to avoid a large bulk carrier three miles or so ahead.

What is very important indeed is an understanding of what closest point of approach (CPA) can be regarded as safe. This parameter is often shrouded in mystery, with the requirement that it must be a 'safe distance'. The reason for this is, of course, because there is no universal parameter, as circumstances can vary and change.

But what is a safe distance? In the open ocean when a vessel is steering by autopilot and a single steering motor with engine running at sea speed, you could regard a minimum CPA of about 2 miles as optimum, with anything closer being a close-quarters situation. In different circumstances, however, a CPA of less than 1 mile might be considered safe, for example, when taking action to pass astern of another vessel which had been on a crossing course and is moving away from your ship. It is the 'what if' scenario that governs safe distance.

You must ask yourself "what happens if my steering develops a problem and the rudder 'freezes' in a position 5° to port or starboard?" How long will it take you to sort this out?

Chapter 5
Navigation Accidents and their Causes

Most experienced navigating officers have experienced exactly this type of problem. My personal wake-up call came in the northwest-going traffic lane of the Dover Strait in a fully-laden Panamax bulk carrier just clearing the Fairy Bank. In my career as an investigator, I have investigated several collisions related to steering gear failure, one of which resulted in the sinking of a medium-sized tanker.

When your vessel is at a state of readiness, with two steering motors in operation and propulsion machinery available for direct manoeuvre in traffic separation schemes and other busy situations, then a CPA of 0.5 miles might be the optimum. Remember this is a distance which amounts to close quarters in an open sea encounter, and it therefore deserves a cautious approach.

This applies to passing distances, but a word of additional caution is necessary. If a vessel in a crossing encounter in open waters shows a CPA of 2 miles with the position ahead of your vessel, then beware. Any number of engine-related mishaps can befall vessels, and these can suddenly result in an apparently safe distance disappearing very quickly. You might want to make that clearing distance greater if it is safe to do so, and you will certainly need to manage the situation carefully.

Use of ARPA in collision avoidance

Second to your eyesight and the ability to observe visually what is happening around your vessel, ARPA is a most useful aid to collision avoidance. However, you must understand its limitations. Consider a recurring factor in collisions between vessels in open waters. This is an obsessive belief held by some that a small CPA margin reassuringly presented on the ARPA, for example less than 0.5 miles, indicates there is no risk of collision. This is most emphatically not the case. Small CPA margins can be dangerously misleading, and what appears numerically as sufficient distance may turn out to be alarmingly close.

An ARPA radar confidently telling you there is a CPA of 0.15 miles from an approaching vessel is actually saying there is going to be a passing distance of 280 metres, which might be just over the length of your vessel. If your vessel is making 18 kt then you will cover that distance in 30 seconds.

Use of AIS in collision avoidance

AIS is another extremely valuable aid to navigation and collision avoidance because it enables the navigator to identify other vessels in the immediate sea area. In most respects, AIS can be viewed as a more sophisticated extension of the ARPA in that it provides more accurate position data for vessels in the vicinity of your ship, together with identification, course, speed and other information. It is of additional assistance when keeping a lookout.

However, exactly the same words of warning apply to AIS systems as do to CPA. The same caution must be applied to the time to closest point of approach (TCPA) information as it does to ARPA. You should also exercise caution because many small vessels are not fitted with AIS so, as with all aids, over-reliance can be dangerous. AIS will provide you with the luxury of more information, but it will not tell you what to do.

Action to avoid collision – Rule 8

The most important requirement is that action taken to avoid collision is positive and readily apparent to another vessel. Alteration of course is usually the most effective way to avoid a close-quarters situation. The key to this is to alter course 'big' and in good time, ensuring there is a large change of heading so there is no mistaking what you are doing.

If you are altering course for collision avoidance under the Rules, then you should make a bold turn of not less than 20° from your original course, or more, if the circumstances require it. Small alterations of course should be avoided, as these can easily pass unobserved. Small alterations of course are often the cause of confusion and misunderstanding on board an observing vessel, and frequently result in collisions. Remember you can reduce the size of your excursion from the course line when the passing manoeuvre has been established and is unambiguous.

Just think about it! If you are walking towards someone in a pedestrian underpass, it makes good sense to make an early choice of which side you intend to pass on and take steps in good time. If you don't do this then there is a good chance you will end up nose-to-nose and saying sorry.

When overtaking, the risks are clear, and the watchkeeper of the vessel which is overtaking should understand the dangers of what might happen if the vessel being overtaken makes a sudden change of course. Such a change might be made by a navigator who is unaware that there is a vessel coming up astern. You should allow a generous passing distance when overtaking to avoid getting caught too close. Consider the long-range traffic and navigational situation, such as potential course changes, and make allowances.

The use of the engines to avoid collisions is a further option. However, it is very important to remember that most vessels manoeuvre slowly, speed changes can be slow to materialise and, importantly, to be detected.

The trial manoeuvre facility is a feature of most integrated navigation systems. This can be a valuable aid but must be used with caution. The experienced navigator has a brain which is very capable of assessing a situation, considering risk and determining a solution, quite often entirely on the basis of visual appraisal. Navigators have a very significant advantage over computer aids – their ability to apply 'weight' to any threat presented. For example, if making a first (or second) alteration for a vessel that is behaving in an unpredictable manner.

Navigators need not be bound by fixed parameters of CPA, and may not require recommendations, which might only serve as a distraction. In most cases, collision avoidance is best kept simple. When taking avoidance action, navigators should be fully aware of their vessel's situation, position and the proximity of any navigational hazards in the area.

It is vitally important that navigators develop and enhance their skills of assessment and do not become over-reliant on electronic aids.

Safe speed – Rule 6

A vessel's speed dictates the time available to assess a situation, and, if necessary, to decide upon action to be taken, to execute a manoeuvre and then, to determine whether the action taken has been effective and that the risk no longer exists.

The key when considering safe speed is to act with caution. Visibility is the most common factor governing speed, but the density of shipping traffic is also very important. Navigators should be aware that turning characteristics at slow speed can be enhanced by increasing engine revolutions to increase the water flow across the rudder. Heading can be controlled very easily at slow speeds, but precise control cannot be exercised if the vessel is obliged, through developing circumstances, to lose headway quickly.

For obvious reasons, it is best therefore to slow down well before arriving off a busy pilotage area or an anchorage.

In some busy harbours and inshore areas, there are bright lights, both ashore and on anchored vessels, which can interfere with the lookout task. Here again it is a matter of exercising caution and providing more time to completely understand the situation, and the way to do this is by control of speed.

Environmental effects can also dictate action. There may be a need to exercise caution and reduce speed but, equally, it might be necessary to maintain speed for the purposes of manoeuvrability.

If you are navigating a fast vessel, which means a vessel moving at a speed in excess of 20 kt, then you should always look for opportunities to alter your course so as to avoid traffic situations altogether. Speed is an enormous advantage in many situations, but you should not arrive at close quarters in a multi-vessel situation when an alteration made at a distance of 10 miles would have allowed you to pass clear of a particular concentration of traffic. On a fast-moving vessel you must think ahead.

Managing collision avoidance – Rule 8 (d)

As important as any action that you take is maintaining a close watch to ensure that your manoeuvres are successful in removing the collision risk. Keep the vessel that you have taken steps to avoid under close observation, and check to see that your action is achieving the desired result. You must, however, be prepared to act again if things are not turning out as you had intended.

You must be prepared for any eventuality, particularly the possibility that the vessel you are avoiding alters course in your direction, perhaps by turning to port when you have turned to starboard.

Chapter 5
A rough guide to collision avoidance

The guidance in situations where things are going wrong is relatively simple. In most cases you should stick to your initial direction of alteration. Remember it can take many seconds, perhaps a minute or more, to reverse a direction of turn. If you maintain your original turn direction, the rate of turn will increase to the maximum. This is where your knowledge of the manoeuvring card may provide some reassurance.

If a situation appears to be developing in an alarming manner and distances are closing within your safety margins, do not be afraid to put the helm over away from danger and take a round turn. All experienced navigators have done this more than once in their careers. However, be sure you check the sea area all around beforehand!

It is always good policy to have a means of attracting attention at hand in the wheelhouse. The signal lamp is the traditional tool for this purpose, and the best bridge layouts ensure the lamp is close to hand at a bridge wing door.

Know and understand Rule 17 – action by stand-on vessel

Rule 17 is a rule which is most frequently misunderstood and misinterpreted by navigators. In the first instance, while a give-way vessel as defined by Rule 15 in a crossing situation shall keep out of the way as required by Rule 16, the other vessel is required to stand-on and keep its course and speed. Importantly, this does mean quite literally what it says.

There is, however, a contingency permitted by the rules. This is the provision of Rule 17 a(ii) wherein a stand-on vessel is permitted to take action by its manoeuvre alone as soon as it becomes apparent to it that the vessel required to keep out of the way is not taking appropriate action in compliance with the Rules. This contingency was a sensible evolution from previous regulations which, without meaning to, actually encouraged vessels to proceed to a position where collision could not be avoided by the give-way vessel alone before action was taken. Modern day navigators, faced with the poor standards of seamanship that we are, on occasions, unfortunate enough to witness, will immediately recognise that such a requirement carries far too much risk.

Under the rules as applied today, the stand-on vessel can act at an earlier stage and need not proceed worryingly into close quarters while 'standing on'. This early action to avoid reaching unnecessary close quarters should logically come after the navigator standing on has studied the give-way vessel for a period of time, and has serious doubts as regards its intentions. There will be an assessment based on a "this is as close as I am prepared to let you get" basis. There may be a give-way vessel which seems intent upon cutting things too fine, or a vessel which appears to be taking no action whatsoever, perhaps because it has not even detected your presence.

The requirement of Rule 17 a(ii) is quite separate from Rule 17 (b), which applies at a much later stage when collision cannot be avoided by the give-way vessel alone. At this point, the stand-on vessel must take action as best it can.

To deal with such circumstances and to set your watch circle or contingency parameter, it once again helps to have knowledge of your vessel's turning circle characteristics. However, it is your CPA parameter that should be applied and, in open waters, a good starting point is the distance of 2 miles referred to earlier.

Navigating in anchorages and port approaches

These areas can be very busy and potentially hazardous with vessels speeding up or slowing down which means that assessing risk of collision can difficult. The key requirement when inbound is to slow down early and proceed with caution. On departure, try to choose a route to open water which is not crowded with waiting vessels. It is by far best to choose the quickest and easiest route to the open sea and then, when in open waters, start passage in the direction required for the next port.

Traffic separation schemes

The theory of traffic separation is that vessels move together in the direction for navigation on the same course, that is, they set down by the direction of the particular traffic separation scheme. The reality can be quite different, and there are numerous quite ordinary technical reasons for this.

The advice is go with the flow, and avoid approaching other vessels on narrow converging courses. There will always be an opportunity to come back to a required course line either at an alteration point within a scheme or at the exit to a scheme. A key requirement is to be flexible as regards courses and way points in a passage plan. The passage plan should not be definitive where way points are concerned, unless there is a particular navigational hazard to be negotiated and passed with caution at a minimum distance. Navigators should have the freedom to make small departures from passage plans as they integrate navigation of vessels with their collision avoidance management.

If you are overtaking another vessel in a traffic separation scheme and you do not have a great speed advantage, then you should provide a wide berth. Overtaking another vessel on the starboard side (your port side) is the preferred position, but this is not always possible.

Use of VHF for collision avoidance

This is a controversial topic, and there is contradiction between the practices of different maritime authorities. The result, perhaps unsurprisingly, is that navigators are falling into the habit of managing collision avoidance by use of VHF communications. This can be fraught with danger and here is why.

The Colregs are quite specific on determining whether risk of collision exists, and in ensuring that action to avoid collision is taken under well-defined guidelines. These are set out in the opening paragraph of Rule 8, where it says:

Chapter 5
A rough guide to collision avoidance

Any action to avoid collision shall be taken in accordance with the Rules of this part and shall, if the circumstances of the case admit, be positive, made in ample time and with due regard to the observance of good seamanship.

By the words *Rules of this part,* navigators should understand that the requirement is to follow the avoidance strategy dictated in sections II and III of the Colregs. A problem is that VHF exchanges encourage less formal, convenience-based strategies which can be quite contrary to the rules. If such measures are successful, then you will probably hear nothing more about it, but if not, you can expect to be very severely criticised.

Further, time can be the enemy of a good navigator. When risk of collision has been determined there will, more often than not, be limited time, or perhaps no time at all, to engage in an exchange on the VHF with another vessel.

Under threat of collision:
- Establish clear and coherent contact
- Be sure of the other ship's identity, even if you identified it using AIS
- Agree a mutual language.

All of this will be time consuming and during this preliminary exchange the distance between vessels will be closing. If the exchange is not meticulously carried out there is a real risk of misunderstanding and confusion and this has resulted in some spectacular collisions.

Another equally important reason to reject VHF exchange is that it encourages action which is not as positive as it should be. Vessels may close from distance and shape to pass with a small CPA. It would be better to taking a broad avoidance manoeuvre to show a clear intention then closing to pass at a safe distance, as defined by the circumstances.

Other vessels in the sea area may not be following your VHF exchange, particularly if, as you should, you have switched to a recognised ship-to-ship channel and are not broadcasting on Channel 16. This can only adversely affect the third party appreciation of a traffic situation, in particular where encounters are being resolved without compliance with the predictable pattern set down by the Colregs.

VHF-guided manoeuvres have a recognised place in collision avoidance, and are ideal for pilotage and confined areas, particularly rivers, canals and narrow channels, where experienced local pilots and navigators can engage with each other. The use of VHF for collision avoidance is not a recommended practice in open waters.

The technology of the future

We are now living with the 'future', as it was viewed 15 years ago, and the new generation of navigation aids – combined radar, ECDIS and AIS displays – are now commonplace. Experienced navigators know that these have the potential to become a distraction when dealing with a simple collision avoidance manoeuvre. It must be understood that there comes a stage where too much information interferes with thought processes.

Chapter 5
Navigation Accidents and their Causes

In the future guidance systems will be created that are capable of directing a safe path for a vessel through shipping traffic by analysing tracks. Another guidance system using repulsive forces to keep vessels away from each other is also being considered. It is not immediately clear how these concepts can be integrated with the Colregs, which are very much dependent upon human logic and apply to vessels of all sizes. The regulations as they exist today provide the mariner with the scope to exercise judgement and apply safety margins.

Route direction

An up-and-coming navigational concept is the provision of a route for a vessel on a prescribed voyage. The primary advantage marketed to charterers by route providers is the ability to save fuel and, depending upon the level of service, there are claims of additional benefits to Masters and owners alike. In its packaged form, it sounds wonderful, but the warning is contained in the very smallest of print. It says that if your route, predetermined by someone sitting in a nice warm and cosy office, has the unfortunate outcome of putting your vessel aground, then you are guaranteed that there will only be one person taking the blame, and it is certainly not the person sitting in the office. There has been at least one very serious grounding already as a result of a misunderstanding over a so-called route.

It remains to be seen whether or not it is a good idea to have large numbers of vessels following the same, most fuel-efficient route, particularly given the accuracy of GPS nowadays. It may, for example, prove a significant test of everyone's understanding of Rules 8 and 14.

In addition, while technology is in the pipeline, life at sea will always tend to move at the pace of the majority. Transferring collision avoidance decisions from experienced human navigators to computers may be some distance away yet.

On a final note concerning technology, have you checked that the AIS data transmitted from your vessel is correct, particularly the ship dimensions and the position of the GPS antenna?

Chapter 6

Pilotage

By Captain Richard Wild

Status and role of the pilot

The status of the pilot can vary from one country to another. Usually the Master retains final responsibility for the ship, but delegates the conduct of navigation, or con, to the pilot. The Masters should always remember that pilots join ships to assist and become temporary members of the bridge team, regardless of whether the pilot is considered to be an advisor or assumes a different level of responsibility. Pilots should not be seen to be doing everything on their own, unsupported by Master and crew.

Boarding and landing pilots

Ships are vulnerable when manoeuvring to board or land the pilot; they are often in close proximity to other vessels and in shallow water. The pilot launch will normally give the ship a recommended course and speed so that the pilot can embark safely. It is important to note that the recommended course is not an instruction; Masters must position their ship so that the recommended course can be achieved and establish that the ship will be able to remain on that course until the pilot boards. This could take a few minutes, but it could be longer if the ladder has to be adjusted. It is not uncommon for the ladder to be either too long or too short. It can be useful to have a selection of short weighted lines of known lengths (2m, for example) that can be attached to the bottom of the ladder. This will assist in getting the ladder set at the correct length.

When checking the boarding course and required sea-room, Masters must take into account other vessels, navigational dangers, the current and wind. After the pilot has boarded, both the launch and the officer at the ladder should inform the bridge that the pilot is safely on board the ship. Pilots will usually (but not always) expect the ship to resume the original course or a course towards any channel or fairway, before they reach the bridge. As pilots proceed to the bridge, they will often gain an impression about the ship and how it is managed; first impressions are important and this also applies to how pilots greet Masters.

Before pilots disembark, they will usually bring the ship to an appropriate course and speed, as advised by the launch. It is important for pilots to inform Masters of the course required, details of other traffic, current direction, speed and any reporting points, before they leave the ship. If this information is not provided, the Master must ask for it.

Chapter 6
Navigation Accidents and their Causes

Full details of pilot boarding and landing arrangements can be found in Chapter 5 of the SOLAS Convention. The IMO has also adopted *Recommendation on Pilot transfer arrangements* (resolution A.889 (21) and approved MSC/Circ.568/Rev.1: *Required Boarding Arrangement for Pilots*. An illustration of the requirements can be found at www.impahq.org

The Master/pilot exchange

The Master/pilot exchange (MPX) underpins every act of pilotage. Numerous incidents have occurred all over the world where one of the root causes identified was an inadequate exchange of information between the Master and pilot. Pilots should engage with Masters and share details of their intended passage plan. They should ensure that the Master knows what is going to happen, where it will happen and when. If Masters do not feel happy about anything, they must ask for further information, regardless of differences in experience and culture.

It is vital that Masters and pilots take this process seriously. If the exchange is poor, there will be little or no team-work. This means that pilots will be operating in isolation. If pilots then make a mistake, they will be the weak link and there could be a rapid escalation to an incident.

To sum up:
- No full exchange = no shared plan
- No planning = no monitoring
- No monitoring = no interventions.

Interventions by the Master or OOW are the onboard safety mechanism which can prevent an accident. There are others, such as shore assistance, and this is discussed later, but the ship's best defence against an incident is the bridge team.

Too often, the exchange is not undertaken seriously or completely. There are two important points to consider which pilots frequently observe. First, when pilots arrive on the bridge of the ship, they cannot be expected to take the conduct, or 'con', immediately. Second, documents which are designed to assist the exchange are often pre-completed, contain errors or there is an apparent reluctance to spend time completing the process, but the pilot's signature is requested on these documents almost immediately. In many cases the message is clear: the ship wants the pilot to take charge as soon as possible and, because the documentation is part of the safety management system, it must be completed and filed.

On the other hand, pilots can appear to be in a hurry and unwilling to talk to Masters and officers. Indeed, three surveys revealed that Masters and pilots report very different opinions of the MPX. This proves that there is an element of mutual distrust, and a lack of awareness and teamwork.

Documentation used to assist the MPX can vary considerably, or might not even exist. Sometimes the document is of an administrative nature for billing purposes; such documents do not assist with the pilotage at all. There are recommended documents at Annexes A1 and A2 of the *International Chamber of Shipping Guide to Passage Planning*, 2007.

Masters should always ensure pilots tell them:
- How any manoeuvres are to be undertaken
- How challenging turns will be made
- Intended passage speed
- Important points in the passage
    - Small under-keel clearances
    - Difficult turns
    - Opposing traffic
    - Tugs required and how they will be used.

If Masters are unsure about any of this basic information, they must ask for it.

Teamwork and culture

Every member of the bridge team must consider themselves to be an equally important member of that team, regardless of their rank. Everyone in the bridge team has an equal voice, but different responsibilities: Masters, officers and others should remember this at all times. Junior officers can find themselves delegated with tasks which do not seem to be important, so it is the responsibility of the Master and pilot to ensure that all team members feel valued.

On a ship, teamwork is required to undertake many tasks, and the bridge is no different. The worst example of teamwork failing is when the pilot is doing everything and is not supported at all. This can be because pilots don't share their plan, or made it clear that they would take over without any assistance.

An example of this is an occasion when a loaded tanker had to abort its inwards passage and was required to turn around and go to an anchorage. The pilot indicated to the Master where the ship could anchor, but did not discuss the best place to turn around and how it would be done. The Master was unable to intervene because he didn't know the pilot's intentions. As a result, the ship collided with a berthed ship.

Many pilots will report that they often board ships where it is quite clear that the Master wants to hand over the ship as quickly as possible. This might be because Masters want to do something else, such as talk to the ship's agent. However, there are many cases where it is obvious that Masters are not happy with the situation at the pilot boarding area. The best way to change this is to plan the boarding manoeuvre properly, and for the bridge team to work together better.

Language and communication

All members of the bridge team must conduct themselves professionally and, above all, say if they do not understand something. There is a tendency for individuals to simply say 'yes' or to report that a task has been completed when it hasn't. This can be disastrous. To earn the respect of the Master and pilot, it is important that proper and honest answers are always given to questions. It is also important that if an instruction is not understood, clarification is asked for immediately. Likewise, those giving instructions and commands must ensure that they speak slowly and clearly. Any unusual requirements, such as mooring arrangements, should be discussed well in advance. Face to face discussions are best and if possible these should take place on the bridge with the pilot.

Wherever possible, important communications should always be shared in a common language. English should be spoken if there is any conflict or difficulty. Internal ship communications, or communications between the pilot and tugs or shore authorities, may be conducted in a common language. However, if there are any problems such as defects or difficulties, these must be shared immediately between the Master and pilot. For example, when the ship is mooring or making fast a tug and the crew are having difficulty with the heaving line, the pilot needs to know.

Tugs and towage

The correct use of tugs is a vital part of manoeuvring ships while a pilot is on board. If there is a lack of understanding, incorrect procedures or a breakdown in communication, there is a real risk that the ship could sustain damage or ground.

Three basic points regarding tugs are: how many are required, where they will meet the ship (or be released from duty) and how they are to be made fast.

The port or berth operator may specify the number of tugs that are required. Typical examples are oil and gas berths. However, pilots can and should be able to demonstrate the number of tugs that are required. Typical reasons are strong wind, under-keel clearance issues and where there is a need for an escort tug to negotiate a turn or slow the ship down. All vessels, but particularly vessels with large wind areas such as container vessels, should work out the windage area in square metres. This will enable Masters and pilots to work out the total bollard pull required, taking into account any thrusters. Various methods exist, but the simplest method can be found in The Nautical Institute publication *The Shiphandler's Guide*. Where there are no port requirements, pilots can experience disputes with Masters who are under pressure to reduce tug use. Windage calculations are a good way to avoid such disputes, and Masters should familiarise themselves accordingly.

Where tugs are to be made fast, or let go, is important. Masters must ensure that members of the crew are standing by to make the tug fast. When a tug is let go and dismissed Masters must be certain that the ship can safely proceed to sea without any further assistance.

Making fast and letting go tugs can be hazardous. The ship can be vulnerable because it must slow down and might not be able to make a large alteration of course until the tug is made fast. The tug may also be vulnerable, particularly if it must make fast in a bow-to-bow configuration with a fine-lined ship such as a container vessel.

When letting the tug go, the crew must follow the instructions from the pilot and tug crew. The tug might have to re-position itself before the tug line is let go; if the line is let go too early, it could be caught into the tug propellers. Tug lines should always be carefully lowered, following instructions from the tug crew. Finally, when assessing how many tugs are required, the Master should be told the type of tugs that will be used and their bollard pull so this can be compared with the safe working load (SWL) of the bitts that are to be used.

Remote pilotage or shore assistance

In some ports, Masters may receive instructions from shore about the course the ship should steer and the speed. These instructions will typically come from a vessel traffic services (VTS) centre, but could also be provided from a pilot mother ship. A VTS operator or a pilot may give the instructions. Radar and AIS will be used to monitor the ship's progress to aid this service.

Provision of these types of service varies all over the world and a definitive list, together with the legal status, is beyond the scope of this chapter. However, remote pilotage and shore assistance are not the same thing. In most countries, it has been determined that pilotage only takes place on the bridge of a ship. Therefore a pilot in a remote operations centre is only able to provide advice. If pilots are unable to board ships because of rough weather, they are unable to give instructions, such as courses to steer, from a pilot launch.

Other countries may appear to take a different view on this, and give instructions to the ship, particularly in rough weather. This could be called remote pilotage. However, the ship should not be given direct instructions. Only the course it should make good (over the ground) in order to avoid immediate danger or to reach a safe waypoint is acceptable. The ship must determine how the course that must be made good is achieved. This type of service is shore assistance.

Regardless of whether the ship receives direct instructions or advice, Masters must always check that the ship will be safe. Masters must always remember that the control centre may be using radar that is some distance away from the ship and not in the best position. In addition, the effect of current and wind on the ship can only be determined on the ship itself. Finally, there will always be inaccuracies in the image in the control centre, caused by time and processing.

Voluntary and compulsory pilotage

Unless Masters have a Pilotage Exemption Certificate (PEC), taking a pilot is compulsory in most ports. However, in some areas of the world, and normally beyond port limits, non-compulsory pilots can be taken. One example of this is the English Channel and North Sea, a large area that includes the approaches to many busy European ports such as Le Havre, Felixstowe, Antwerp, Rotterdam, Bremerhaven and Hamburg.

These voluntary or deep sea pilots have normally served as ships' Masters and are familiar with the sea areas but not the port. Pilots must undertake an examination before they are licensed to work.

Use of these pilots can be advantageous for a number of reasons: they have good local knowledge; experience of navigating in areas of traffic congestion; the ability to make all required VHF reports; assist with timing issues and assist with fatigue and stress management.

Keeping records and checklists

While keeping records is important, it should not take up so much time that an individual tasked with completing a bell book is, for example, unable to assist the pilot and bridge team. An illustration of this is when a ship ran aground in poor visibility, while approaching a berth. The OOW had to make a large number of entries into the bell book so was not able to monitor the ship's position and provide information about the ship's speed to the pilot.

Voyage data recorders are now common on many ships. These should be used in a positive way. Masters and pilots should stand in a position so that as the voyage progresses discussions including MPX are recorded together with all helm and engine orders.

Checklists are an important part of any SMS although, too often, they are seen as the only tool that will ensure that the ship does not have an incident. This is not the case. Checklists are developed from good practice guides, procedures and previous experience. But checklists cannot cover every circumstance and should only be seen as the start point for a safe port entry or departure. Masters and pilots must always ask themselves if there are any unusual circumstances that might not be covered by the normal checklists. And there is nothing worse than pre-completed checklists presented to pilots for signature.

Future trends

Predicting the future is always difficult. However, in the short to medium term, technological advances will continue to dominate the ship's bridge. For example, projecting information such as course, speed or even channel limits onto the bridge window is now possible. Radar has remained unchanged for many years, but advances using 4G signals are expected. ECDIS is now a reasonably mature technology, but its implementation has not been helped by the proliferation of manufacturers and models. In the future, more standardised displays can be expected.

However, not all technological advances are seen on the bridge. Already, the cost of fuel and emission control policies are having an impact on ship propulsion, vessel loading and design. Ships are being retrofitted with different bulbous bows and high-efficiency propellers. Some vessels are being altered to carry more cargo, but engine power has not increased and can even be reduced. Generally, ship's engines are designed to run as efficiently as possible, and the smallest engine that is suitable is installed. This means that the ship may have a limited manoeuvring range and is under-powered. This can have an impact in high winds and affect turning ability. Masters and pilots must be aware of this.

Ocean Breeze (image: ISU)

Chapter 7

Under-keel clearance

By Dr Tim Gourlay

In this chapter, we will be looking at the topic of under-keel clearance (UKC) and what mariners can do to avoid ships grounding in shallow navigation channels.

The first thing to be considered is the static under-keel clearance (Static UKC). This is the available water depth, including tide, minus the draught of the ship. It is shown in Figure 7.1.

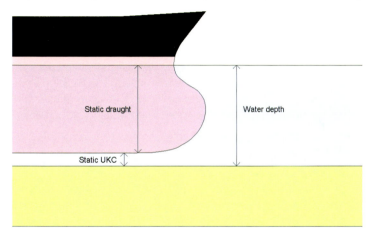

Figure 7.1 Static UKC: static draught is taken to be the maximum of fore and aft draughts

If the ship was drifting along the waterway with no vertical movement, this would be its minimum clearance from the seabed. Calculation of the static UKC requires detailed knowledge of:
- Chart datum depths through the waterway
- Tide height at each time and location through the waterway
- Ship forward and aft draughts on entering the waterway. These may be based on readings taken on departure from the previous port, allowing for fuel consumption and the seawater density in the waterway.

For ships in shallow water, grounding risk is not just about the water depth and the ship draught. The ship also has its own vertical motions, which need to be taken into account. Squat, wave-induced motions and heel each cause parts of the ship to move closer to the seabed, and the combined effect of all of these needs to be included.

Chapter 7
Navigation Accidents and their Causes

This chapter begins with some case studies of ships that have run aground unexpectedly, despite the water depth being larger than the draught (positive static UKC). We then look at how to allow for vertical motions of the ship and how to combine each of the allowances into an overall UKC management.

Ship grounding case studies

It is useful for the mariner to examine case studies of ship groundings that have occurred wholly within navigation channels, when the depth was greater than the static draught, so that the grounding may be attributed to squat, heel and/or wave-induced motions. Several groundings over the years have been blamed on squat, but close inspection reveals that there is often significant uncertainty about the exact circumstances at the time of grounding. Determining the static UKC at time of grounding depends on exact knowledge of the bathymetry, ship's track, local tide height and actual ship draughts. It is rare that all of these factors are accurately known. For example, in the days before AIS ship tracking, we normally couldn't be sure of a ship's actual track, and in estuarine channels, salinity may not have been accurately accounted for. Also, without nearby tide gauge measurements, the tide may only have been estimated. Early groundings for which squat was probably an important factor include the *Wellpark* at La Plata Roads in 1977 (Ferguson et al, 1982) and the *River Embley* in the Torres Strait in 1987 (ATSB 1988). However, for these cases it is difficult to determine the static UKC accurately.

Case studies

Queen Elizabeth 2 (QE2), Vineyard Sound, USA, August 1992. A grounding attributed to ship squat and coarse-grid surveying. The severity of this grounding was caused by the high speed (25 kt) at the time of impact. Evidence of grounding was found on two previously unsurveyed boulders: Red Rock I, where the static UKC was 0.8 m; and Red Rock II, where the Static UKC was 0.6m. Any modern squat formula would predict that the *QE2* would run aground under these conditions.

Another grounding in which squat was an important factor was that of the oil tanker *Sea Empress*, on approach to Milford Haven, February 1996. In that case, the minimum static UKC was 0.0m, and at 10 kt the bow squat was calculated to be approximately 0.75m.

Jody F Millennium, Gisborne, New Zealand, February, 2002. Grounding primarily attributed to wave-induced rolling which occurred when the log carrier was making an emergency departure from the port. The ship was forced to leave the harbour, having parted eight shore mooring lines from wave-induced surging in the harbour. The ship was on a south-westerly heading, with wind and swell approximately from the south. No wave buoy was present at the time of the grounding, but a wave buoy further south had recorded significant wave heights of up to 8m. The static UKC at the time was approximately 1.5m. A large wave caused the ship to roll to starboard, then to port, and ground on the port beam ends. The ship then rolled back to starboard and grounded heavily on the starboard beam ends, at which time it slewed to starboard, drifted out of the channel and ran hard aground.

Capella Voyager, Marsden Point, New Zealand, April 2003. Grounding attributed to wave-induced heave and pitch in the approach. The ship's track, bathymetry, tide height and ship draughts were known sufficiently accurately to determine a minimum static UKC of 2.6m. This case represents possibly the largest static UKC at which a ship has run aground. *Capella Voyager's* speed at the time of grounding was a maximum of 6 kt, at which the bow squat was calculated to be approximately 0.48m. This case demonstrates the large wave-induced motions that can occur in long-period swells. No wave buoy was present at the time of the grounding, but the wave conditions were estimated to be a long-period swell with significant wave height in the order of 3-4m from the east (stern quartering seas).

On contacting the rocky seabed, the hull of the *Capella Voyager* was pierced just forward of the collision bulkhead. Shell plating was also creased over a length of 21m, from the tip of the bulbous bow to just forward of the collision bulkhead. The *Capella Voyager* grounding brought into focus the issues of communication and decision-making between the Harbour Master, pilot and ship's Master. In this case, the Master believed the conditions to be unsafe, and the pilotage company believed that the conditions were safe; ultimately the transit went ahead and the ship ran aground.

Eastern Honor, Marsden Point, New Zealand, July 2003. A similar grounding to that of the *Capella Voyager* occurred three months later at the same location, where the *Eastern Honor* ran aground with 1.5m Static UKC. These two incidents illustrate the need for accurate real-time wave data and detailed UKC modelling in swell-affected ports.

Desh Rakshak, Port Philip, Australia, January 2006. Grounding attributed to a combination of squat and wave-induced motions. It highlighted the fact that ship squat is dependent on speed through the water, rather than speed over the ground. At the time of the grounding, the ship was travelling at 8 kt over the ground and 13 kt through the water. The minimum Static UKC was 1.2m, so that at this speed the ship was predicted to run aground due to the effect of squat alone. However, there was also an appreciable swell running. No wave buoy was present at the time of the grounding, but the ship was reported to be rolling with an amplitude of around 5° in the following seas. The rocky seabed caused extensive damage to the shell plating and tank framing, and pierced the hull ahead of the fore peak bulkhead.

Squat

Squat is the downward vertical movement and change in trim that is caused by a ship's own wave pattern when travelling at forward speed. Figure 7.2 shows the nature of the changed wave pattern along the ship at varying speeds.

60 | Chapter 7
Navigation Accidents and their Causes

Figure 7.2 Wave pattern around ro-ro cargo ship *Seagard* at 6.5m draught in 14.5m water depth. Above: speed 10 kt; Below: speed 20 kt. Images courtesy of John Clandillon-Baker

To understand the squat phenomenon, it is easiest to change our frame of reference to one in which the ship is fixed, and the water is streaming past it. This is what is done for model ship testing in a re-circulating water tunnel. In this frame of reference, the flow is steady and we can use the Bernoulli equation which says that when the flow speed is high, the pressure is low, and *vice versa*.

Here is one explanation often given for the squat effect. Because of the small under-keel clearance, mass conservation requires that water is forced at high speed beneath the ship. This high flow speed reduces the pressure underneath the hull and pulls the ship downwards.

This explanation is only partly true. In deep water, water flows underneath and around the ship. In shallow water, there is nothing forcing the water beneath the ship – it just goes around the sides instead. But it is true that the presence of the ship and seabed accelerates the flow, decreasing the pressure on the ship and pulling it downwards.

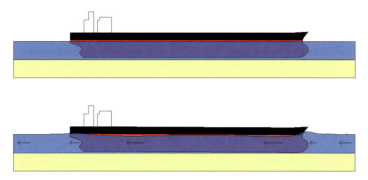

Figure 7.3 Panamax bulk carrier with draught 12.2m in 14.0m water depth. Above: ship at rest; Below: ship travelling at 12 kt. Arrows show changing flow velocities around the hull in ship-fixed reference frame. Note lowered free surface at forward and aft shoulders, where flow speed is highest. Ship is shown in squatted position, with bow close to grounding

Figure 7.3 shows the local flow velocities and free surface height around a Panamax bulk carrier travelling at 12 kt in shallow water. At the bow, there is a stagnation zone with low flow speeds and high pressure. This high pressure translates into an elevated free surface. At the forward and aft shoulders, the water is accelerated, causing low pressure and a lowered free surface.

Overall, the presence of the ship causes a generally accelerated flow and lowered free surface. The ship then sinks downwards with this lowered free surface. It will also change its trim depending on the relative magnitudes of the wave troughs at the forward and aft shoulders. High-block-coefficient hulls such as bulk carriers and tankers have large hull curvature at the forward shoulders, resulting in a very low pressure there and resulting bow-down dynamic trim.

The flow acceleration described above is increased further in laterally-restricted water such as dredged channels and canals. Figure 7.4 shows how dredged channels, and especially canals, create a blockage effect which increases the flow velocities and pressure changes around the hull.

Chapter 7
Navigation Accidents and their Causes

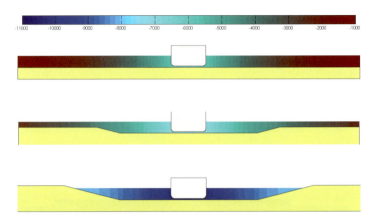

Figure 7.4 Pressure field at midships for a container ship travelling at 12 kt in open water, a dredged channel or a canal. Colours show pressure in Pascals above hydrostatic pressure

Formula to predict squat in constant depth

First calculate the Froude depth number F_h

$$F_h = \frac{U}{\sqrt{9.81h}}$$

Here U is speed through the water in metres/second, that is 0.514 times the speed in knots, h is the water depth (including tide) in metres. The factor 9.81 is the earth's gravitational acceleration in metres/second^2. For cargo ships in shallow water, the Froude depth number is normally in the range 0.3-0.6.

Now find the ship's displaced volume ∇. In sea water, this is the total displacement in tonnes, divided by 1.025. Onboard a ship, the displacement may be found from the *Trim and Stability Book* at the correct draught, or alternatively from the *Trim and Stability Report* calculated for the present ship condition. Without this information, the displaced volume may be estimated using a block coefficient C_B of 0.82 for bulk carriers and tankers, 0.75 for LNG carriers, or 0.65 for container ships. The displaced volume is then estimated using:

$$\nabla = C_B L_{PP} BT$$

Here L_{PP} is the ship's length between perpendiculars, B is the beam and T is the draught. In open water, squat may be estimated using the ICORELS formula. This has been shown to give results accurate to within around 20%, over a range of bulk carrier, tanker and container ship hulls. It is one of the formulae recommended by PIANC. It should not be used for high-speed applications when the Froude depth number is greater than 0.7. The formula is:

$$S_{max}=2.4\frac{\nabla}{L_{PP}^2}\frac{F_h}{\sqrt{1-F_h}}$$

This gives the maximum squat over the length of the ship. For bulk carriers and tankers, it will tend to be at the bow, while for LNG carriers and container ships it may be either at the bow or the stern.

As an example, consider the case study of the tanker *Desh Rakshak* (see page 59) at the time of grounding. At a speed through the water of 13 kt and water depth 12.7m, the Froude depth number is 0.60. With L_{PP}=234.0m, *B* = 42.0m and *T*=11.5m, its displaced volume is approximately 93,000m³. According to the ICORELS formula, the maximum sinkage is then 1.82m.

In dredged channels, squat will be 5-10% larger than the above formula. In canals, it will be approximately double the formula, depending on how confined the canal is.

Some high-speed displacement ships, such as frigates, destroyers, cruise ships and catamaran ferries, can travel at Froude depth numbers well above 0.7, which is the approximate limit for the squat formula. In this case, it is helpful to know the maximum sinkage, which occurs at a Froude depth number of 0.9-1.0. In this 'trans-critical' speed range, there is large midship sinkage and large stern-down trim, meaning that the maximum sinkage occurs at the stern, where the propeller may be vulnerable. The ro-ro cargo ship shown in Figure 7.2 has a Froude depth number of around 0.85, and illustrates the wave patterns observed in the trans-critical speed range.

At trans-critical speeds, the maximum sinkage can be estimated (Gourlay 2006) using

$$S_{max}=1.5\frac{\nabla}{L_{PP}h}$$

As an example, a frigate of displacement 4,000 tonnes and length between perpendiculars of 125m, travelling in water depth 15m (including tide), would need to allow for a maximum sinkage of approximately 3.2m. This maximum sinkage would occur at a Froude depth number just beneath 1.0, which corresponds to a speed of 23.6 kt in this case.

Roll, heave and pitch

The case study of *Capella Voyager* on page 59 demonstrates that in long-period swells, grounding can occur at large static UKC values (2.6m in that case). The important motion components are roll, heave and pitch, each of which produce vertical motions of the ship.

Roll and heel

The effect of ship rolling on UKC can be analysed in a simple way. Figure 7.5 shows an end view of the *Jody F Millennium* at the time of grounding. Its beam is 26.0m; draught 9.5m and the water depth is 11.0m.

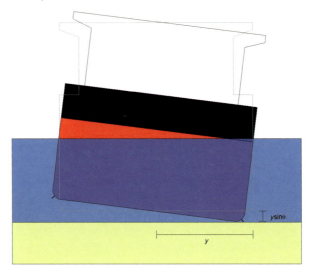

Figure 7.5 Effect of roll on UKC; example showing simplified end view of the *Jody F Millennium* at time of grounding

The distance from the ship's centreline to the turn of the bilge, or tip of the bilge keel, is **y**. The ship is rolling with amplitude **θ°**. This causes the beam ends to move vertically, by an amount **y sin (θ)**. In Figure 7.5, a roll angle of 7° is shown, in which case the *Jody F Millennium's* bilge keels grounded. Since roll is routinely measured onboard ship, this simple calculation allows an estimate of the decreased UKC when under way due to rolling.

The method may also be used for wind heel, which is primarily important for car carriers, container ships and LNG carriers. It may also be used for heel due to turning, which is primarily important for container ships. Standard methods exist for predicting heel angle due to wind or turning, see for example PIANC (2014).

In order to predict which wave conditions are likely to produce large rolling, it is helpful to know the ship's natural roll period. This is the period that a ship would roll at, if pulled over to one side and then released. Large roll angles occur when the wave period is close to the ship's natural roll period. Using SI units (metres, seconds), the natural roll period T_ϕ is approximately

$$T_\phi = 0.8 \frac{B}{\sqrt{GM}}$$

GM is the transverse metacentric height, corrected for free surface effects. As an example, consider a Panamax bulk carrier with beam 32m. When fully loaded, it has a GM of 2.0m, and natural roll period of around 18 seconds. Therefore in long-period swell (for example, where the swell period is greater than 15 seconds) we would expect large roll angles to occur. In ballast condition, if it has a GM of 8.0m, the natural roll period is 9 seconds. This is much more in line with typical wave periods, so we would expect large roll angles to occur in normal ocean conditions. Noting that the fully-loaded condition is normally the critical case for UKC, wave-induced rolling is normally an issue only for ports that are exposed to long-period swell.

Heave and pitch

Waves in ship navigation channels, particularly head seas or following seas, produce pitching motions which can make the bow or stern vulnerable to grounding. The ship's centre of gravity also moves up and down, which is the heave motion. Depending on the phasing between heave and pitch, the bow or stern may have larger vertical movement.

Heave and pitch motions in shallow water are quite different from those in deep water. Waves of the same period are shorter and travel more slowly in shallow water than in deep water. The ship's motions are also damped more by the presence of the seabed. All of this means that ship motions tend to be smaller in shallow water than in deep water, which is good news for UKC.

In order to understand ship wave-induced motions, we need firstly to understand the waves. Consider a navigation channel with depth 14m (including tide). Wind waves with period 5s would have a length of 38m. These will cause large motions of small vessels, but the wave loads will tend to cancel out over the length of a cargo ship. Therefore cargo ships are largely unaffected by wind waves.

Consider now a swell wave with period 12s. This has a wavelength of 134m, which is now comparable to the length of cargo ships. Figure 7.6 shows a Panamax bulk carrier of L_{PP}=180 m heading into these waves. We can see that the aft half of the ship may be mostly in a wave crest, while the forward half of the ship is mostly in a wave trough. The result is a bow-down pitch angle, which can lead to grounding.

Figure 7.6 Panamax bulk carrier in head seas. Wave period is 12 seconds and wave height is 4m. Draught is 12.2m, water depth is 14m and bow is close to grounding

A real ocean wave pattern comprises a range of different wave periods and possibly directions. The combined effect is a near-random variation of wave height with time. Of special importance is the occurrence of wave 'sets'; typically one to four waves in a row, with height much larger than the average. Therefore a statistical approach is necessary, which aims to determine wave-induced motion allowances based on an acceptable long-term grounding risk. These statistical approaches are discussed in PIANC (2014).

UKC management in practice

For ships transiting port approach channels:
a. Masters must be satisfied that there is sufficient UKC according their ship-specific calculations AND
b. Ports must be satisfied that there is sufficient UKC according to their port-specific calculations.

In practice, the port-specific calculations tend to govern the transits, as these should include detailed modelling of the local conditions. In essence, channel bathymetry and metocean conditions vary more between ports than hull shapes vary between ships. However, Masters' knowledge of the behaviour of their own ships and their UKC judgement, are an important part of the decision-making process. The grounding of *Capella Voyager* (see page 59) is an important illustration of this. The good judgement of the Master is especially important in 'unusual' environmental conditions, where standard UKC guidelines may fail.

For each ship, it is useful for the crew to have a squat table readily available on the bridge. This squat table should be ship-specific, as, for example, wide-beam ships squat more than narrow-beam ships. The squat table can be developed at full load draught and using a typical minimum operating static UKC. It may be based on model tests or numerical modelling.

The squat table will typically be developed for wide channels or open water. It should be recognised that the actual bathymetry also has an important effect. For example, in canals such as Suez or Panama, the squat may be around double the open-water value. Ideally, dedicated numerical modelling should be done for very confined channels or canals.

Dynamic trim is very ship-specific, and the ship's crew should develop an understanding of the way their ship trims with speed.

UKC management techniques for ports typically fall into three categories:

- Fixed minimum static UKC

This method is ideal for port approach channels that are protected from ocean swells, and where simplicity is required. For example, the high-volume port of Kwai Chung in Hong Kong uses a fixed minimum static UKC of 15% of the draught (HKPA 2012). Despite the simplicity of the method, extensive validation has been undertaken to confirm its validity across a wide range of ships and operating conditions, including full-scale validation trials (see Figure 7.7).

Chapter 7 | 67
Under-keel clearance

Figure 7.7 Measuring container ship vertical motions at the port of Kwai Chung, using real-time-kinematic GPS. Image courtesy of CMST

- UKC tables

For ports where metocean conditions (especially swell) are variable, but simplicity is important, paper or spreadsheet UKC tables may be used. These give the required static UKC for each ship class as a function of the most important input variables (typically swell height and period). The required static UKC includes the effect of squat, heel due to wind and turning, wave-induced motions and safety/manoeuvrability allowance.

- UKC software

Due to the limited number of inputs available, the previous methods may be overly conservative in some cases. UKC software allows more inputs and more specific UKC modelling. It includes allowances for squat, heel and wave-induced motions, and facilitates the in-built inclusion of tide and swell predictions and real-time measurements. Examples of UKC software include:
- *CADET/EMOGS*, developed by US Government agencies
- *DUKC®*, developed by OMC International, Australia
- *HARAP*, developed by the Port of Rotterdam
- *MetOcean UKC*, developed by MetOcean Solutions, New Zealand
- *ProToel*, developed by the University of Ghent, Belgium.

Chapter 7
Navigation Accidents and their Causes

Figure 7.8 Screenshot of *MetOcean UKC* software for assessing ship transit safety at the Port of Taranaki, New Zealand. Image courtesy of MetOcean Solutions

The increased usage of UKC software internationally should help to improve safety and efficiency of ships in navigation channels. However, it also brings new risks to mariners, including the challenge of trusting software of which they have little knowledge.

UKC software is necessarily complex, and for commercial reasons the specifics of some UKC software are kept closely guarded. For this reason, it may seem to mariners that it is a mysterious package, with workings they do not understand. Therefore we need to encourage transparency in UKC software. Ideally, methods should follow international guidelines such as PIANC (2014). In this way, mariners can learn the methods and limitations of UKC software, and trust it accordingly. Ideally, pilots for each port should also be involved with the software development from an early stage, so that they can feel comfortable with and know the limitations of the software.

Unusual conditions

UKC software has limitations for unusual ship and environmental conditions, when a mariner's experience may provide better judgement, or at least a common-sense check on the software. Examples include:

- Container ships normally have little dynamic trim, but some have large bow-down dynamic trim, making the bow vulnerable to grounding. This effect is ship-specific and unlikely to be predicted using UKC software. It is then for Masters to know their own ships, and make allowances accordingly.

- During a new swell front, wave height and period can both increase very rapidly, making conditions dangerous for ship grounding. UKC software, using either forecast or real-time wave data, will be subject to error in this case, and the mariner will need to exercise caution when using UKC software.
- External factors, such as heel due to tugs, may not be included in UKC software and may need to be separately assessed.
- Decreased salinity during heavy river outflows, or large tidal residuals due to cyclonic or tsunami activity, are other examples of unusual conditions which might not be adequately accounted for in UKC software.

Conclusions

For ships in shallow water, allowances have to be made for vertical motions due to squat, heel and wave-induced motions, in order to ensure safety from grounding. Such allowances may be included in an overall UKC allowance, or be used in UKC tables or UKC software. The mariner should understand the conditions under which a large Static UKC is needed, such as high speed or long-period swell, and exercise caution in these conditions.

References

Ferguson, A M, Seren, D B, McGregor, R C (1982). 'Experimental investigation of a grounding on a shoaling sandbank'. RINA *Transactions*, Vol 124.

Gourlay, T P (2006). 'A simple method for predicting the maximum squat of a high-speed displacement ship'. *Marine Technology and SNAME News*, Vol 43, No 3.

Hong Kong Pilots' Association (2012). Berthing guidelines. Endorsed by Pilotage Advisory Committee, Marine Department, HKSAR.

PIANC (2014). Harbour approach channels – design guidelines. PIANC Report No 121.

CMA CMG Florida (image: MAIB)

Chapter 8

Anchoring

By Captain Nadeem Anwar

Incidents at anchor have been the cause of many significant insurance claims and remain a concern in the industry. Vessels at anchor are at risk if anchoring is not properly planned and conducted correctly. Risks will not be the same at every anchorage and some key risk elements will be more important than others, depending on the nature of the location. Professional and competent watchkeeping while at anchor can prevent a difficult situation occurring and a proficient anchor watch can help manage events before they start to pose a threat to the vessel, crew or the environment.

This chapter sets out to provide some useful reminders of steps which should be taken to avoid accidents from anchoring operations.

Anchoring risks

It is imperative that a risk assessment is performed by Masters and their bridge teams before a vessel proceeds to anchor. All too often, accidents have occurred because of a failure to carry out this essential task.

The main anchoring risks which mariners should consider include:
- Dragging
- Grounding
- Collision
- Fouling of anchor and cable with submerged objects or moorings
- Damage to submarine cables and pipelines
- Windlass failure
- Inclement weather
- Piracy activities.

Additional risks may arise if a vessel is exposed to other hazards, which might include:
- Water depth
- Strong current and tidal streams
- Adverse weather
- Lee shore
- Seabed composition
- Inadequate sea room for swinging
- Inadequate sea room for heaving anchor and manoeuvring out
- Proximity to other vessels anchored nearby and navigational hazards

- Proximity to vessels passing or transiting the anchorage location
- Incorrect laying of ground tackle and scope of cable
- Main engine and auxiliary problems.

Of course these factors are not exhaustive and will not all feature at any one time. However, it is crucial to ensure that any and all factors which might affect a vessel at anchor are considered and measures put in place to ensure that incidents can be avoided.

Let us move on to the planning stage where more detail is discussed.

Planning for anchoring

To secure a safe anchoring, Masters and bridge teams have to take a number of factors into account in a 'what if' risk-based approach and allow for any changes that can happen after the planning stage. Forethought at this stage can save the need for a difficult situation while at anchor.

The choice of a safe anchorage should consider these factors at the very least:
- Designated or recommended anchorage position
- Commercial limits, including restraints imposed by coastal states or port authorities
- Sub-surface hazards and obstructions
- Composition and profile of the seabed
- Water depth
- Prevailing or characteristic sea state, wind and current or tidal stream
- Anchor location approach speeds (very important if location has strong tidal effects)
- Shelter available
- Duration of stay at anchor
- Other vessels at anchor and adequate room for anchoring
- Proximity of services which may be required while at anchor
- Density of traffic and proximity to approach channels
- Security considerations
- State of engines and anchor equipment
- Design limitations of anchoring machinery and equipment.

Planning for anchoring and time at anchor needs to be part of the overall passage plan and should involve the bridge team and engine room staff too. Key elements which need to be taken into account will be the Master's experience and that of other senior officers; guidelines and procedures set out in the vessel's SMS and guidance from all available publications including Sailing Directions, commercial port guides and any recommendations from port authorities concerning the features of the anchorage.

Once the location is selected or approved by port control, if applicable, the next phase is to plan the actual anchoring operation. Experienced Masters will be aware of the importance of laying the ground tackle to ensure the best holding of the anchor.

The bridge and anchor teams must be given a detailed briefing about bringing the ship to anchor. As mentioned earlier, Masters must take a risk-based approach when planning for anchoring with every risk assessed and managed carefully. Providing all deck officers with training and experience in anchoring operations is particularly useful and will ensure that experience is shared to the mutual benefit of the whole bridge team.

Prevention of dragging anchor

The risk of dragging anchor is greatest where the holding ground is poor. There are many examples of dragging when such risks have been ignored. Anchoring on a seabed with rocks and sand is a typical example and once dragging begins, unless detected promptly, a vessel can quickly run into serious difficulties.

When considering an anchorage, the nature of the seabed is very relevant to the holding power of the anchor. For example, mud and clay are very cohesive and sand is less so. There are some very good seamanship publications which give guidance on these factors. Sailing Directions and commercial port guides also provide relevant advice on holding ground at anchorages and give details of particular features of anchorages such as prevailing winds and shelter. This advice, coupled with onboard experience, provides the best indicators for a suitable anchorage based on the suitability of holding ground. Prudent mariners will know that rocky areas should be avoided. While they may provide good holding ground at times, anchors can be damaged or become fouled in the rocks and recovery may not be possible, or extremely difficult at best.

Another factor is the slope of the seabed which increases holding power if the load is in the ascending direction and reduces holding if the load is in the descending direction. A ridge between the anchor and vessel improves holding.

Holding power of anchor and scope of cable

Masters should be aware of the characteristics of the ship's anchor. They should also be aware of the guidelines for holding power in various conditions and if these change, that the stated holding power may not apply.

If the scope of anchor chain is inadequate the vessel may drag anchor, even in average conditions. The depth of water is the key factor when deciding the scope of the anchor cable (one shackle of cable = 27.5 metres). There are a variety of calculation methods and guiding principles regarding what is considered an adequate scope of cable. A common guiding principle is that the scope of anchor cable should be about $1.5 \times \sqrt{}$ (depth in metres) when using a high-holding power anchor, and increased if the swinging room allows to compensate for a poor holding ground, significant fetch with high winds, large swell or an adverse bottom contour. Another rule of thumb is to pay out 6-10 times the water depth. There is also the sensible precaution that one should always keep some chain in the locker in case of the need to range the scope out a bit more if the weather changes.

Chapter 8
Navigation Accidents and their Causes

There is a risk of having an inadequate scope of chain. An ideal situation is to have about one-third to half of the cable in water to lie horizontally along the seabed and the remainder of the cable to lie in a catenary. The amount of cable lying horizontally along the seabed adds to the holding power of the anchor.

> **Case study**
>
> *Ropax 1*, Algeciras Bay, December 2008. The dragging of the anchor was clearly due to the use of insufficient scope of anchor cable combined with worsening weather conditions, according to the MAIB report. Readers would do well to study this report and heed the warnings it contains.

Should the cable lift above the seabed by even a small angle, the holding power of the anchor system can be reduced significantly. For example, an angle of 5° to horizontal could reduce holding power by 25%.

Condition of vessel

Vessels at anchor will behave differently depending on whether they are fully loaded, in a normal or light ballast condition or part loaded. Clearly, in ballast condition, the windage area of the vessel is increased, resulting in higher loading on the anchor and cable. Additionally there is likely to be forward slamming when the vessel is in light ballast condition when seas are from ahead. The laden condition of the vessel should always be taken into account when planning anchoring.

The vessel's load condition is also relevant when there is a need to clear an anchorage or to change location if heavy weather is forecast or increases without warning. A classic example of failure to clear an anchorage in a timely manner is the incident involving the bulk carrier *Pasha Bulker* (also see Chapter 11, page 104).

> **Case study**
>
> *Pasha Bulker*, near Newcastle, NSW, Australia, June 2007 (see Chapter 11, page 104). The ship was carrying inadequate ballast and a combination of various factors, not least the failure to heed bad weather warnings, resulted in its grounding with a subsequent complex and costly salvage operation.

Currents and tidal streams

The risk of dragging anchor in a strong current or tidal stream is exacerbated further when a vessel is loaded. Masters and OOW should always be vigilant in observing direction and rates and how they affect the vessel at anchor. Remember that the flow of water against the submerged hull places the main load on the anchor cable. While at anchor, vessels tend to ride stemming the current or tidal stream. The surface area exposed to current or tidal stream depends on the beam and draught of the vessel: the deeper the draught and wider the beam, then the greater the load on anchor systems. All this should be known by Masters and the knowledge passed on to junior deck officers. Involvement of the bridge team in anchoring operations is essential both at the planning stage, execution of anchoring operations and maintaining anchor watches. Having all watch officers aware of the particular risks involved for each anchorage is essential to avoid incidents.

The usual scope of cable typically applies for currents or tidal streams up to 3 kt. If the water flow is greater, then additional scope of cable needs to be deployed. Similarly if the vessel has a wider beam, deeper draught or both, the scope of the cable would have to be increased even in normal flow conditions. Where a larger vessel is exposed to excessive flow while at anchor, the scope of cable has to be increased accordingly. Generally every 2 kt of additional flow rate would require an extra shackle of cable. Masters and deck officers should be aware of these features and ensure they are included in anchor planning and when keeping the anchor watch.

Wind

Obviously a vessel which is in ballast or is high sided is more affected by wind and the windage area of the vessel would have a similar effect in terms of load on the vessel and the anchor cable. Where wind is strong, the vessel would ride to the anchor stemming the wind. In such a case the load on the vessel depends on freeboard, beam and the height of the superstructure. A prudent and vigilant Master and bridge watch will keep a close eye on prevailing and forecast weather. With increasing winds or forecast to increase, extending the scope of cable in a timely manner might well be the sensible thing to do. Masters should always be aware of how a particular vessel behaves in weather conditions. The design of the vessel, for instance whether the accommodation is forward or aft, will have a bearing on how the vessel will lie at anchor.

Wind, tidal stream or current at angle to each other

In certain parts of the world where both wind and tidal stream are strong the vessel may ride the anchor cable in an unusual manner. The vessel may be stemming the tidal stream, but would be pushed sideways due to the wind. In such a case, and given insufficient scope of chain, the anchor system can experience excessive loads, giving rise to the real risk of dragging anchor. Again, Masters and the bridge watch should always pay particular attention to the effects of the prevailing environmental conditions and how they are affecting, or may affect, the vessel at anchor. To be proactive is eminently better than taking a reactive approach.

Excessive yawing

Excessive yaw while at anchor can expose the anchor cable to excessive or snatched loads, an excessive load can cause a vessel to drag its anchor. This may happen where wind and tidal stream are from different directions. Another condition is where heavy swell is being experienced by the anchored vessel from a direction different from that it is riding the anchor cable. In extreme circumstances, the Master may consider deploying the second anchor or heaving anchor and laying off until suitable conditions prevail. It is important that these features and their effects are understood by the bridge watch and that Masters are proactive in taking action in a timely manner to overcome any adverse effects. Having the main engine ready and anticipating contingency requirements will avoid the risk of a serious incident.

Swell or wave height

The effect of sea and swell on the anchor needs to be fully understood. When the vessel experiences heavy weather with large waves and/or swell, it pitches. In the case of a ballasted vessel, it would pitch more than in the loaded condition. Where pitching is excessive, it lifts the bow significantly. As the bow lifts high enough, additional cable is lifted off the seabed. This reduces the holding capacity due to snatched loading at that time and the vessel may drag its anchor.

Case study

Willy, Plymouth Sound, UK, January 2002. The grounding of the tanker is an example of excessive loading on anchor cable when pitching in large waves. In such circumstances, additional scope of chain is prudent. This report is highly recommended reading.

Exposed location

A vessel is at risk of dragging when anchored at an exposed location. This risk can be managed by selecting a location which shelters the vessel from the elements, especially when the conditions are expected to exceed certain limits. An alternative anchorage location minimises the risk of grounding and in addition highlights a number of key issues.

Case study

Stena Alegra, Karlskrona, October 2013. The grounding of this ro-ro passenger ferry brought to the fore the need for a sheltered location in line with wind and weather conditions. The vessel was anchored off Karlskrona, Sweden, and dragged anchor in 76 kt winds and beyond the design capability of the anchoring system. A failure to identify the consequential risks was highlighted in the report. Readers are encouraged to review the MAIB report which underlines the need to make careful risk assessments in anchoring.

Changing conditions should be monitored continually and Masters should not hesitate to weigh anchor and move to a sheltered location or to open sea, if deemed necessary. Master's Standing Orders should emphasise the need for constant monitoring of weather conditions and to call the Master if ever in any doubt.

Risk of grounding, or impact with seabed

If the risk of dragging can be managed, then the threat of grounding can be substantially reduced. If the vessel does start to drag anchor the availability of propulsion systems and the operational status of mooring and anchoring equipment are key preventative factors.

Due to changes in direction of the tidal flow and the wind direction, vessels can swing around an anchor position several times a day. Sufficient sea room must be taken into account in all directions to allow for a safe swinging circle. The time of swing is always a concern as the anchor flukes will change direction in the seabed. At this stage the anchor may dislodge itself from the seabed and may be caused to drag. In addition to the risk of dragging, the swing may bring the vessel nearer to other vessels at anchor or hazards in the vicinity. When selecting an anchorage, the sea room and swinging circle are key factors to consider. If other vessels anchor too close then communication should be made to alert them of the risks of contact during swinging (see also the section covering risk of collision with other vessels below). The grounding of *Stena Alegra* (page 76) highlighted the importance of anchoring at a safe distance from navigational hazards especially in light of prevailing, or forecast, wind conditions.

Minimum depth

Appropriate depth ensures a safe under-keel clearance (UKC) at all times while the vessel is at anchor (see Chapter 7 page 57). However, based on a risk assessment, it should be understood that an offshore wind or significant high pressure could reduce the effective depth to below the charted depth. Therefore at low water the UKC may be reduced significantly, or the vessel may even touch bottom. Rolling, pitching or heaving motions of the vessel under these circumstances of reduced under-keel clearance could also lead to contact with the seabed. Chapter 7 of this publication provides comprehensive information about UKC considerations and is an excellent reference.

Risk of collision with other vessels, or impact with fixed or floating objects

Approach channels and traffic

Hydrographic authorities usually mark charts 'anchoring prohibited in port approaches'. Even if no prohibition is marked, Masters need to assess risks from traffic flow and approaching vessels to ensure they anchor in areas clear of routine or known traffic flow.

Chapter 8
Navigation Accidents and their Causes

Bridge teams need to continuously review all risks from moving traffic. Nearby vessels should be monitored in case they start dragging their anchors and if this is detected, the observations should be communicated as a matter of priority. A dragging anchor is not only a danger to your vessel but to others too. The dragging of anchor of the ro-ro *Ropax 1* at anchor off Algeciras in close proximity to a monobuoy and a fish-farm resulted in the vessel making contact with the monobuoy (see page 74). Pre-arrival plans must take into account the subsea and surface hazards. However, on arrival, planning must consider the position of other vessels already at anchor and the available sea room should other vessels come to anchor.

Risk of damage to anchor and cable, and vessel's mooring equipment

Case study
Young Lady, off Teesport, UK, June 2007. When the tanker started to drag anchor, an attempt was made to heave up the anchor but the vessel suffered a windlass failure which was reaching its design limitations in the prevailing conditions.

Masters and officers should be familiar with such design limitations and feed these into the risk assessment when planning to anchor and during the stay at anchor with a careful eye on weather conditions. Accidents can be avoided with such knowledge.

Maximum depth

A vessel's windlass capacity is one of the limits that restrict the maximum depth in which it can anchor. When a windlass remains operational and power remains available it can continue to heave horizontally, subject to load on the cable; but the situation changes with vertical lift. Again, knowledge of the anchoring equipment and machinery and the design limitations is crucial in anchoring operations.

Risk of fouling the anchor and cable with submerged objects

Anchoring prohibited

Anchoring may be prohibited in an area for a number of reasons. This could be in order to avoid blocking the approach channel or because of the presence of buoy moorings. Other prohibitions may be because of dumped explosives or the presence of old cables or sub-sea polluted areas.

Chapter 8
Anchoring

Risk of damage to submarine cables and pipelines

Cables and pipelines

Submarine cable and pipelines areas exist around many ports in close proximity to, or in, approach channels. Vessels should avoid anchoring close to pipelines and/or submarine cables. Many maritime administrations have severe penalties for anchoring in close proximity to these assets and local regulations must be consulted. There have been many occasions when vessels have fouled underwater power cables which could have been avoided with proper inspection of the navigation chart. In June 2007, although initially anchored well clear of any submarine pipelines, the *Young Lady* started to drag anchor off Teesport UK and – having suffered a windlass failure – dragged on to a submarine pipeline (see page 78).

Before anchoring, it is essential to make proper reference to relevant nautical publications to ensure that the intended anchoring position is not within a prohibited area or in very close proximity when any dragging might result in encroaching upon such an area.

Weather risks

The onset or approach of storms must be taken seriously and, after a careful navigational risk assessment, vessels should put to open sea at the earliest opportunity. The Master of *Pasha Bulker* (see page 74 this chapter and Chapter 11 pages 104-105) had received instructions from the port authority to proceed to sea to avoid a tropical storm, but did not do so in time. When the ship did leave the anchorage, it was too late. The delay in taking action was a primary reason for the grounding of the vessel. Changing weather conditions leave ships vulnerable to dragging or grounding.

Anchor watch

Masters and officers should be aware of the need for a vigilant anchor watch. However, it is as well to provide some reminders of essential considerations and requirements. In addition to careful planning and risk assessment, appropriate procedures must be properly applied at all stages. Watchkeeping instructions for the time at anchor need to be absolutely clear from Masters. Immediately on anchoring, the following duties must be performed:
- Plot the anchor position
- Plot the swinging circle based on anchor position
- Note the limits of the bearings expected, especially abeam and any transits abeam
- Note the ranges for different possible headings ahead and astern to fixed objects
- Set up anchor watch alarm on GPS and ECDIS
- Set up radars to monitor position and other vessel movements
- Mark the cable on deck to check slippage
- Note and plot the position of other vessels anchored nearby

Chapter 8
Navigation Accidents and their Causes

- Set VHF to required channels to monitor port traffic, VTS and others as required for warning of other vessel movements.

While maintaining the anchor watch, it is essential to plot positions regularly using the bearings and distances taken. In changing conditions, the position should be checked even more frequently.

Monitoring

Even at the safest anchorage, vessels may drag anchor if the prevailing conditions change or if anchoring has not been managed properly. Do not use GPS alone to ensure position at anchor is maintained within the swinging circle. Conventional means of monitoring position at anchor should always be utilised. With this in mind, an anchor position should be chosen where effective monitoring is possible through marks on the beam for checking bearings and distances ahead or astern for checking ranges.

Security

Ships at anchor can be vulnerable to attacks by pirates and thieves. Port radio channels should be monitored and all company and port security measures maintained at all times.

Master's Instructions, SMS and ISPS

It is preferable that Masters designate specific teams to anchor stations during watch periods with each team member briefed on their specific duties in advance. Masters need to set detailed standing orders and write specific instructions for watch officers to manage changing situations. The SMS manual should provide clear advice and instructions for anchoring operations. The ISPS should also be heeded too.

Contingency planning

Masters must ensure that contingency plans are in place while at anchor to allow for changes in weather or other conditions. On the slightest indication of a change in position or noises or vibrations coming from the anchor cable, the watchkeepers should immediately advise the Master and contingency plans should be initiated. Remember, normal loads and holding capacities do not apply when the anchor starts to drag. The rate of drift during the dragging stage can be alarming.

If the anchor starts to drag, there are a variety of options available to bridge teams and Masters. The exact approach taken will depend upon the many things that have been taken into account when planning the anchoring. Masters should empower their OOW to use the engines to prevent vessels from dragging on to other vessels and hazards.

Readiness of main engine

The quickest and easiest way to control dragging is through prompt use of the main engine. The engine must remain ready for use while the vessel is at anchor, especially in circumstances where environmental conditions or traffic is likely to cause concern. Engine repairs at anchor should always be viewed with the prevailing circumstances and likely need of main engines. Many incidents at anchor have been caused because the main engine was not available for immediate use. Some ports have designated anchorages for engine maintenance and repairs and port control should be advised accordingly.

Actions on dragging

Masters and bridge teams may have to consider the following if the vessel starts to drag anchor: use of main engine to ease load on cable and to arrest dragging; increase the scope of the cable; let go the second anchor if deemed necessary; and if all this fails, weigh the anchor or anchors and steam out to open sea and ride out the bad weather or seek an alternative anchorage, depending on what the cause of dragging was.

In any event, before the onset of inclement weather, preparations should be in place to take appropriate action. Clearing the anchorage for open sea is very often the safest option to take. If there is a problem in weighing the anchor, and as a last resort, slipping the chain might be the only option to avoid being caught in a disastrous situation.

Conclusions and the future

Anchoring, like many other activities at sea can be fraught with danger. However, with proper planning, risk assessment, monitoring and by implementing proper procedures such dangers can be minimised. Situation awareness through vigilant anchor watchkeeping will always play a key role in anchoring operations. Efficient bridge team management will be pivotal throughout. Chapters 2 and 3 of this book address passage planning and bridge resource management respectively, both of which are key factors in anchoring operations.

Shore monitoring of anchored vessels by vessel traffic services does provide an additional safety tool and monitoring system. However, such services are not widespread and it will take many years before this is likely. Navigation systems are continually improving and it is not beyond belief that anchor monitoring will be more refined in the years ahead to assist Masters and OOWs. However, for now, and the foreseeable future, there is every reason why anchoring operations need to be treated with utmost caution and care to ensure that accidents can be avoided.

Further reading

Standard Club safety bulletin on anchoring procedures, October 2008, www.nauticalplatform.org/docs/files/Standard%20Club%20Anchoring%20procedures.pdf

(image: Danny Cornelissen)

Chapter 9

Electronics: Some friendly advice on bridge work

By Professor Thomas Porathe

Things have changed considerably on ships' bridges since my great grandfather sailed as chief officer of the three masted barque *Hilda* on the Sweden-South America trade at the end of the 19th century. Then the instrumentation in the chart house on the quarterdeck was sparse to say the least: a sextant, a chronometer, a barometer, charts and nautical tables and the patent log streaming from the aft rail. These were the most important navigation aids. The rest had to do with experience and rules of thumb. Knowledge was general – standard currents, standard movements of weather systems in different places of the globe; approaching a landfall after days of overcast skies was something to be seriously concerned about.

Today the situation is pretty much the opposite. Today we know most things: the whereabouts and movements of weather in real time, the position of own and other vessels down to tenths of metre, the whereabouts of land and reefs. By contrast, problems today are related to information overload, complexity, of understanding automation and, from time to time, outright boredom.

This chapter will try to give some advice to steer through these problems based on some accident investigation reports. These are real accidents, recommendations that follow from different authorities and some examples I have encountered in my research.

Familiarisation

The chances are that you will come onboard a new ship with only a couple of hours to go until it is necessary to cast off the lines for your first watch. On the bridge, you will find the familiar electronics: the radars, the ECDIS, the GNSS, the AIS and VHF units. You will have seen similar units many times before, but this time the manufacturer or the model may be new to you. Let's hope you will not be alone on your first watch.

The IMO and The Nautical Institute ECDIS familiarisation guidance say you should have time for familiarisation, but this is always a challenge on a busy ship. You must take every opportunity to get to know the workings of the equipment as soon as possible. Start with the basics and investigate the specifics of each system. Keep practising as you never know when your proficiency will save you at some time in the future.

Chapter 9
Navigation Accidents and their Causes

> **Case study**
>
> *Sleipner,* Norwegian Sea, November 1999. The high-speed ferry slammed into a rock at 32 kt, subsequently sinking. Sixteen of the 88 passengers and crew onboard died.
>
> The accident happened at night in bad weather. According to the report the two officers on the bridge had been navigating visually using the sector light along the coastal fairway. During a crucial passage, the Master wanted to apply fixed range rings on his radar. His usual posting was as Master of the *Sleipner's* sistership; the brand of radar was different. It took him a few seconds too long before he worked out how to set the fixed rings on this new system. The other officer was busy tuning the gain control to reduce sea clutter on the other radar set. In the meantime, no one was looking out of the bridge window.

Remember from college: two independent navigation methods.

The two officers on the bridge in the *Sleipner* casualty above only used one navigation method: visual. They went up and down the coast every day and knew the coast and the archipelago very well. However, on the approach to the lighthouse they needed to be sure they did not run into the light, something that is difficult to judge in the dark. Therefore, they needed another navigation method: radar. But it takes some time to set up the radar, not much, but more than they had. You, of course, will always hopefully have your radar tuned and set.

Check your settings

Do not assume that radar, ECDIS and other instruments on the bridge are set once and for all. Learn where the settings are and how to operate them. Be aware that the previous watch might have changed settings, or that different weather, sea or approaching coast conditions might need new settings. In one case, a large bulk carrier collided with a pleasure yacht because the OOW did not see the small sailing craft on the radar.

> **Case study**
>
> *Furness Melbourne* and *Riga II*, north of Bowen, Queensland, May 2012. Collision between a bulk carrier and a yacht. The ATSB report stated that when the Captain came on the bridge after the collision he immediately went to the radar and adjusted the gain control to reduce sea clutter; the yacht was easily visible.

It is important to keep training and learn how the radars onboard your ship work. Keep learning and keep practising.

Chapter 9
Electronics: Some friendly advice on bridge work

Case study

Pride of Canterbury, south-east coast of England, January 2008 (see Chapter 4 page 34). The large ferry sought shelter close to land due to severe weather, but grounded on a charted wreck and badly damaged one of her propellers. According to the MAIB report, the OOW had not observed the wreck because it was not displayed on the ECDIS screen due to inappropriate settings.

There is scope for ECDIS to be personalised, allowing staff to hide information. This is useful to de-clutter screens. But that makes it difficult to be sure that all the information needed is displayed. The chances are that you will be taking over a customised display on your next watch. Make sure you know how to set and reset the system to the operating modes that you prefer.

Case study

LT Cortesia, Dover Strait, January 2008. The large container vessel grounded on a sandbank not far from where the *Pride of Canterbury* grounded. Although the vessel had a draught of about 12 m, the safety contour in the ECDIS was set to 30 m. With this setting, all water areas with a depth of 30 m and less were displayed in a blue non-navigable colour rendering the colour warning system of the chart useless. The safety contour also triggers the depth alarm in the ECDIS. The setting of deep contour, shallow contour, safety contour and safety depth can be problematic as the evidence of accidents and incidents shows.

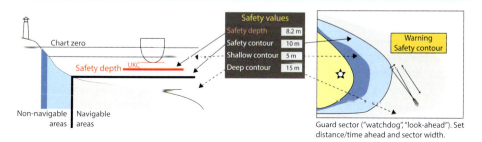

Figure 9.1 Understanding the safety depth and safety contours in ECDIS may be difficult. If set up correctly, they will help you check your route and monitor the voyage. If set up incorrectly, they may trigger alarms that could tempt you to disable vital safety warnings. Settings may look and work differently according to the equipment brand

The set up in Figure 9.1 might look familiar. The safety depth – the ship's draught plus the desired UKC – is the most important setting. In most systems, setting the safety depth will automatically set the safety contour to the first available level below the safety depth. It will also highlight spot soundings which are shallower than your safety depth. If you want, you can set the deep and shallow contours yourself. But remember, it is the

Chapter 9
Navigation Accidents and their Causes

safety contour together with the guard sector settings that will trigger an alarm. If you do not keep these settings updated unnecessary alarms will sound. The chances are that you will then disable the alarms and deprive yourself of a valuable aid to navigation.

The *LT Cortesia* had a draught of 12 m and a UKC of 2 m, so the safety depth should have been set to 14 m. This, in turn, would have activated the next available deeper safety contour, which in this ENC was the 20 m contour. Figure 9.2 shows the visual difference between an ECDIS with a 20 m safety contour (left) and a 30 m safety contour (right). Notice the difference and how danger can suddenly develop.

Figure 9.2 (a) and (b) The Dover Strait with the grounded vessel on the Varne Bank. Left (a), ECDIS with the safety depth not set so in the default 30 m setting. Right (b) with the 20 m setting. Images from MAIB´s reconstruction of the accident at Warsash Maritime Academy

The sandbank is much more conspicuous in the left image. The 30 m safety contour is the default setting if no safety depth is entered, based on an IMO recommendation. Figure 9.2(b) shows that in the default setting the entire SW bound traffic channel is within the non-navigable area. Even so, the 30 m safety contour should have set off an alarm just before the vessel struck the sandbank. Not so. According to the accident report, the depth alarm function was turned off. When the *LT Cortesia* tried to avoid two buoys, mistaken for fishing vessels, it grounded. And it happened again...

Case studies

Ovit, Varne Bank, Dover Strait, September 2013. (Chapter 4 and Chapter 11, pages 34 and 106). The accident to the small chemical tanker was very similar. The safety depth was also set at 30 m – that is, not set at all – although draught was only 7.9 m. As with the *LT Cortesia* the depth alarm function was turned off. But in this case the alarm port had never been configured since the installation of the ECDIS system, so that no alarms could ever be activated. The ship went straight over the sandbank.

Chapter 9
Electronics: Some friendly advice on bridge work

Trans Agila, north of Kalmar, Sweden, November 2012. The container vessel's planned track passed on the wrong side of a light and it grounded in a coastal fairway in the Baltic Sea. The watch officer was alone on the bridge as the lookout had gone down to prepare for embarking the pilot, who was shortly due. The ship's engine room flooded and the vessel was declared a total loss.

How could this happen? After the passage planning, safety checks of the route must be made. Every step must be checked. Alarms should warn of too small UKC. Evidently this was another case of disabled alarms. We have to examine why people want to disable alarms.

Turning off alarms

Alarms are something special. The most frequent complaint from the bridge concerns unnecessary or faulty alarms that cause stress and increase workloads. We all know what a relevant alarm is: it warns us of something we have forgotten or missed. We are thankful for these alarms. But if they warn us of something we already know, are not relevant to our current work or are faulty, then we get annoyed. Bad alarm handling has led to a situation where alarms are either incapacitated by different means, or are acknowledged in a routine fashion with no notice taken of the warnings.

I have spoken to developers of bridge equipment about this, but they claim the alarms are stipulated in performance standards. So it seems that the problem stems from regulations. So the bridge team has to find a way to deal with the issue. Some turn off alarms as much as possible, even to the point of cutting wires and acknowledging alarms automatically without trying to find out what is wrong. This means you are deprived of a warning that might save you one day. The other way is to become proficient in working alarm systems, learn what triggers them and make sure settings are correct so fewer go off unnecessarily.

Learn a lesson from the small container vessel *Godafoss,* where the depth alarm was turned off.

Case study

Godafoss, off Fredrikstad, Norway, February 2011. The vessel left a Norwegian port one evening and passed out through the archipelago with a pilot, the Master and the first officer on the bridge. Just before leaving the archipelago, the boat came to pick up the pilot. The first officer followed the pilot down on deck to help him disembark. The Master was alone on the bridge to take the ship through the final strait, past a couple of beacons. He had done the trip many times before. Moving in the white sector of the lighthouse ahead he was to turn 8° starboard when passing the first red light on his starboard side and pass the lighthouse in green sector on his port side. Instead he, for some reason, continued straight ahead, moving out of the white sector and into red and the vessel grounded a couple of hundred metres in front of the lighthouse.

Chapter 9
Navigation Accidents and their Causes

Figure 9.3 *Godafoss* grounded in the Norwegian archipelago in 2011. Image courtesy of Accident Investigation Board Norway

Pay attention

This does not mean you should spend your watch training and checking settings. Not so. There will be many boring watches when the ship is peacefully steaming an empty ocean and the automatic navigation system is taking care of course keeping. It can then be very tempting to try to multitask to use the time effectively. One of my students, who had returned to college after several years at sea as a watch stander to take a higher degree in maritime logistics, told me about an incident. He had been the OOW and the ship had for hours slowly proceeded in a channel approaching port. He felt very bored and nothing happened, so he took up his iPhone and half-heartedly played a simple game while he looked up every so often to check the traffic. Suddenly, as he looked up, there was a dredger right under his nose. That time the situation was saved.

The OOW of the ferry *Alandsfarjan* was not quite so fortunate as it approached Mariehamn in the Baltic Sea. As the vessel approached the buoyed channel on autopilot, the OOW was checking his service schedule on a personal computer on the bridge. Meanwhile the ship collided with a small island at 13 kt.

These are examples of human error and are results of the boredom that highly automated bridges engender. Humans are not designed to monitor perfectly capable automatons for years on end, just waiting for a mishap or a breakdown that might never come. One of the greatest challenges for those on the bridge in the coming years will be to stay alert on watch. Use the time wisely, use alternative methods of navigation, train and learn. Look at rehearsing some examples.

Chapter 9
Electronics: Some friendly advice on bridge work

Mental simulation

Accidents statistics show human error as a major cause but do not record human recovery. Often human inventiveness saves a situation that might otherwise have led to an accident. One way of keeping alert on long watches is through mental simulation or rehearsal.

Pose some questions. What do you do if the GPS signal were lost right now? How would the equipment react? What warnings would you get? What could you do about it? Engaging yourself in such mental simulation will increase your readiness, and the chances are that you might be better prepared once something does happen.

In one close call the situation was saved because two quick minded officers, one on the bridge of each ship, did the right thing where many others might have stopped thinking.

Case study

Tarnfjord near collision with *Wellamo*, Hastholmen, Stockholm approach, August 1991. The product tanker *Tarnfjord*, loaded with 20,000 tonnes of gasoil, was under way through a complicated Scandinavian archipelago. In a narrow bend in the fairway there was a routine meeting with one of the large ferries trafficking the area – the *Wellamo*. The ferry had nearly 1,000 passengers and crew onboard.

As the tanker applied starboard rudder to negotiate a bend in the fairway, the Master noticed that the rudder instead turned to port and a port turn started a few hundred metres in front of the oncoming ferry. He immediately reversed the engine, but realised he would not be able to prevent the turn, so called the ferry on the VHF saying they had a breakdown on the steering engine and asked for green-to-green meeting. The ferry responded promptly, but by making a starboard 360° turn, and the ships passed each other on parallel courses with 20-30m between.

The accident investigation board calculated that if the action from the ferry had been delayed by 30-60 seconds it would have been impossible to avoid a collision with the ferry, running into the amidships section of the tanker at a right angle. The consequences can only be imagined.

But, mind you, there is a fine line between mental simulation and daydreaming.

Reliance and over reliance

Is it true because it is in colour? The bridge of a modern ship is an impressive experience with all the expensive electronic systems showing you the whereabouts of both your own and most other vessels in the vicinity. Why bother to look out of the bridge window? All these instruments are built according to performance standards and of course you must rely on them. As your experience grows you will find that they really are reliable. But does that mean that they can never fail?

Chapter 9
Navigation Accidents and their Causes

Case study

Royal Majesty, Nantucket, June 1995. Grounded with 1,500 passengers and crew onboard on a sandbank 17 miles off course. The antenna cable to the GPS units had become disconnected and without anybody noticing it the chart system switched to dead reckoning. A classic example of absolute reliance and certainty that the electronics could not be wrong.

The position was now calculated from gyro and speed log but no longer making good for leeway due to wind, current and sea. No one on the bridge team noticed that the input source had changed from GPS to dead reckoning and they continued to sail for more than 24 hours in this mode. The autopilot maintained the course and the OOW could see the own ships symbol move along the planned track on the radar and chart systems. As they approached land in the nightfall they started to get close to buoys and beacons. The OOW checked the chart, and the ship was precisely where it was supposed to be on the course line.

The first buoy on the chart should be on the port side, and looking out the window against the setting sun he could see a buoy at about that position. The characteristics of buoy and light were not checked, however. The next buoy never appeared. But the chart system showed that the vessel was right on track. There were several distant buoys around as the night fell but no light characters were checked; the radar range was never set up to check long distance.

The accident report called it over reliance. Think about that: what is the difference between reliance and over reliance?

In the case of the *Royal Majesty*, the ship was equipped with both GPS and Loran C receivers. All the bridge crew should have compared the readings of the two positioning systems once in a while. Now Loran C is effectively obsolete (but could be replaced by eLoran in some parts of the world). On the other hand, a GNSS outage will not go unnoticed on a bridge. All units using satellite positioning input will alarm as stipulated by the performance standards.

Then you will have to go back to old fashioned navigation without GPS – can you still do it? There is also a nasty thing called jamming. A small, cheap transmitter can be hidden onboard, or brought onboard in a car or lorry that does not want to be tracked. If the jammer is blocking the GPS signal you will at least get an alarm. But there is a signal level where the GPS will not alarm, but instead the position will randomly walk around in the vicinity. What will you do then? You might wonder what is happening.

And so to conclude. The risks of over reliance on electronic navigation aids cannot be stressed too much. It is imperative for watchkeepers to have sound knowledge of equipment and its settings and limitations. It is also essential to monitor the aids frequently and compare with other navigation systems, particularly if in any doubt. Over reliance can, and does lead to accidents including collisions and groundings. Take heed!

Chapter 10

Vessel Traffic Services

By Captain Terry Hughes

There is no single answer to preventing accidents, although some key requirements when navigating in coastal and harbour waterways are good communications and a good working knowledge of all equipment being used, especially an awareness of its limitations. Training and education, both ashore and onboard, are particularly important, especially with regards to on the job competence and continuing professional development.

Governments have been working for generations, at national and international levels, to work towards the safety of shipping in busy traffic lanes and port approaches.

The timeline for the development of these services began in February 1948 when the first port control radar was installed at the entrance to the Isle of Man's harbour. However, it was probably the Port of Liverpool which pioneered vessel traffic services (VTS) when, in the same year, it set up a combined radar and radio station, in order to facilitate the boarding of pilots. Since then, ports and governments have all worked to improve port operations.

In 1951, Long Beach in California established a similar system to facilitate its port operations. Other major ports in Europe quickly followed. Although at this time commercial radar was comparatively new it was now possible, under almost all weather conditions, to observe vessel traffic from the shore. In combination with VHF radio, a traffic surveillance system was achieved, and real-time information exchange between the shore and ships became possible.

IMO and SOLAS

In 1968, The Inter-Governmental Maritime Consultative Organization (IMCO), as the International Maritime Organization (IMO) was then called, published Resolution A.158 – *Radio Advisory Services*. This recommended that governments should consider setting up such services, particularly in ports and oil terminals where noxious or hazardous cargoes are loaded and unloaded, and that for safety reasons, Masters should notify the appropriate authorities of their arrival time as early as possible.

In 1985, the IMO adopted Resolution A.578(14) – *Guidelines for Vessel Traffic Services*. This recognised that the safe, efficient movement of maritime traffic within a VTS service area is dependent upon close cooperation between the vessels and those operating the VTS.

Chapter 10
Navigation Accidents and their Causes

It also recognised that the use of differing VTS procedures in different service areas may confuse Masters of vessels moving from one to another.

In 1997, the IMO adopted Resolution A.857(20) – *Guidelines for Vessel Traffic Services,* which is still in force today. This Resolution includes two important annexes: *Guidelines and Criteria for VTS* and *Guidelines on Recruitment, Qualifications and Training of VTS Operators.*

Although Resolution A.857(20) is the current guideline on VTS, it is SOLAS which is the most important with respect to VTS as it contains compulsory VTS requirements.

The sinking of the *Titanic* on 14 April 1912 after colliding with an iceberg was the catalyst for the adoption in 1914 of the first International SOLAS Convention. In the year 2000, amendments to Chapter V of this Convention were adopted, which included Regulation 11 – *Ship Reporting Systems* and Regulation 12 – *Vessel Traffic Services.* Regulation 12 consists of five paragraphs, one of which states that contracting governments planning and implementing VTS should, wherever possible, follow the A.857(20) guidelines.

What is VTS

VTS is defined by the IMO as a service implemented by a 'competent' authority. It is designed to improve the safety and efficiency of vessel traffic and to protect the environment. The service should also have the capability to interact with the traffic and to respond to traffic situations developing in the VTS area.

In ports, harbours and coastal areas, vessel traffic is managed and regulated by a VTS authority, which provides relevant services to that traffic and to allied services when required. The VTS operation is carried out from a VTS centre, which, generally speaking, is a highly-visible tower very similar to that used by airport traffic control.

If the port does not operate a VTS, there may be a local port service (LPS) instead, only imparting information to a vessel relevant to the particular port itself. An LPS will not interact with vessel traffic outside the port area, nor will it provide any of the services that a VTS does.

The main objectives of VTS are to improve the safety and efficiency of navigation; the safety of life at sea, and the protection of the marine environment and the adjacent shore area, worksites and offshore installations from possible adverse effects of maritime traffic. In addition, VTS helps to improve port operations by easing the safe passage of vessel traffic within the harbour or coastal area. This is achieved by sensoring equipment capable of generating information, communications equipment and comprehensive monitoring of the traffic image, by whatever means are available.

VTS is often likened to air traffic control (ATC), and there are similarities, particularly as both rely heavily on communications. However, while VTS operates in two dimensions and is primarily concerned with managing traffic within territorial waters, ATC operates in three dimensions and ensures spatial clearance in all globally-controlled airspaces.

Chapter 10
Vessel Traffic Services

Although the VTS is able to monitor any movement that a vessel makes, it may not be able to see targets such as, for example, small sailing vessels within close range of a vessel, particularly if the vessel is hiding them in a radar shadow. The VTS does not have any of the instant live shipboard sensors or visual cues such as rudder angle and waterway currents. This means that both the VTS and vessel have limitations with respect to their visual displays and sensors, which is why they must work together to achieve the main goal of safe navigation.

Navigational and safety communications between ship and shore and between vessels must be clear, concise and correct. This is of particular importance in the light of the increasing number of internationally trading vessels with crews speaking many different languages, since the different ways of communicating may cause misunderstandings, leading to dangers to the vessel, the people onboard and the environment. All communications must be diplomatic, not dictatorial. In other words, they need to be polite and informative as well as being without conflict.

In order to try to standardise the language used in communications and to help avoid confusion and error, the IMO published Resolution A.918(22) – *Standard Marine Communication Phrases* (SMCP). Under STCW 1978, as revised 1995, the ability to use and understand SMCP is a requirement for the certification of officers in charge of a navigational watch on ships of 500 gross tonnage or more. A working knowledge of VTS procedures is now included in the STCW Code.

Types of VTS

The type of VTS is largely dependent on the size of the area, the result of carrying out a risk assessment, vessel traffic density, nature of cargoes and passengers being transported, as well as helping to protect valuable man-made and natural assets including bridges, underwater reefs, renewable energy sources and densely populated areas adjacent to a waterway.

There are three main types of VTS: coastal, inland and port or harbour. Coastal type VTS is generally used for surveillance purposes, carried out in sensitive areas where some form of traffic management is required to ensure that vessels, passing through an area, comply with IMO-adopted routeing measures. For example, the English and French VTS authorities cooperate in assisting specific types of vessel transiting the Dover Strait, imparting information when required.

Inland VTS ensures the safe transit of vessels in rivers or estuaries, on their way to a port. For example, in the Thames estuary leading to the Port of London; the Malacca Strait on the approach to Singapore; the Turkish Strait between the Mediterranean and the Black Sea and the archipelagic approaches leading to Turku in Finland and Stockholm in Sweden.

In Europe, vessels using inland waterways may cross the borders of several countries. The operation of VTS in the adjacent areas should be uniform, where appropriate, in order to avoid confusing Masters of inland vessels when crossing VTS areas of different types.

Where two or more governments or competent authorities have a common interest in establishing a VTS in a particular area, they should develop a coordinated VTS on the basis of an agreement between them. International guidelines on VTS in inland waters should follow the IMO guidelines on VTS as closely as possible.

In the USA, one example where the benefits of a public-private partnership work very well is the Los Angeles-Long Beach (LA/LB) VTS, which is jointly operated by the Coast Guard and Marine Exchange of Southern California.

Mandatory ship reporting systems

Under SOLAS (Chapter V Regulation 11) the IMO has approved the establishment of many mandatory ship reporting systems in several major shipping areas of the world. These systems may include traffic separation schemes (TSS) and/or routeing measures (Chapter V Regulation 10). The number of collisions and groundings has often been dramatically reduced in such areas.

In 1967, after much consultation, the world's first voluntary TSS was established in the Dover Strait. This scheme reduced the number of collisions, but as shipping volume increased, so did the accident rate. In 1971, as a result of this increased accident rate, including loss of life, the first mandatory ship reporting system (MSRS) was established there.

Over the years, a number of MSRS have been established, examples of which are the Straits of Dover, Gibraltar and Malacca and the Great Belt in Denmark. The latter was primarily set up to help protect the bridge in the Great Belt (BELTREP). Australia has two major mandatory reporting systems, the Australian Ship Reporting area (AUSREP) and the Great Barrier Reef and Torres Strait Ship Reporting System (REEFREP). AUSREP covers the western and southern approaches to the Australian continent and most of the coastline. Very often, mandatory ship reporting areas are managed from a VTS centre, as is the case of Reef VTS and Great Belt VTS.

When in an MSRS, ships are obliged to give information about themselves, such as their identity and type of cargo carried, to coastal authorities. The authority can then track voyages and communicate with ships as and when required. However, there should be no expectation on the part of vessels transiting such areas and participating in an MSRS, that the authorities concerned will provide any navigational or meteorological information.

The implementation of an MSRS makes it easier to avert hazards, which can be caused by unidentified ships adopting erratic or even dangerous routes, stopping in a traffic lane after sustaining damage, or otherwise behaving in such a manner as to give rise to confusion.

Role and responsibility of VTS

Governments around the world are responsible for the safety of navigation and protection of the marine environment in areas under their jurisdiction. This responsibility is delegated to either their competent or VTS authority, which will have the authority and responsibility for the management, operation and coordination of the VTS, interaction with participating vessels and the safe and effective provision of the service.

Some VTS authorities are managed by government agencies, while others are managed solely by civilian authorities. While rules and regulations may vary from one VTS area to another, VTS and communication procedures should be consistent.

In order to discharge the VTS authority's responsibility and duty of care for the navigational safety of vessel traffic (commercial and non-commercial), the most important functions are those related to:
- Safety of life at sea
- Safety of navigation
- Efficiency of vessel traffic movement
- Protection of the marine environment
- Supporting maritime security
- Supporting law enforcement
- Protection of adjacent communities and infrastructure.

It is important that VTS operations do not encroach upon the Master's responsibility for safe navigation, or disturb the traditional relationship between Master and pilot. The ultimate responsibility for safe navigation of a vessel always remains with the Master, as at no time is the VTS relieving the Master of the responsibility to control the vessel's movement.

The safe navigation of a vessel in a VTS area is all about teamwork. The Master will have a bridge team, and while on board, the pilot should also be integrated into the team. Bridge resource management (BRM) uses all available information and assistance to ensure that navigators make the best possible decisions. The VTS is indirectly part of the bridge team, although it provides its information from an external centre. The information should be available in time for onboard navigational decision-making. BRM training is highly recommended, and falls within the requirements of the STCW Code.

Types of service provided by a VTS

There are three types of service provided by a VTS to vessel traffic: information service, navigational assistance service and traffic organisation service. A VTS may provide any one or all three of these services in its area of responsibility. The type of VTS and services provided by an authority may be promoted in nautical publications, on navigational charts and on the internet.

Information service (INS)

An information service ensures that essential information is available in time for onboard navigational decision-making. This service is provided by broadcasting information at fixed times and intervals or when deemed necessary by the VTS or at the request of a vessel, and may include reports on the position, identity and intentions of other traffic; waterway conditions; weather; hazards or any other factors that may influence the vessel's transit.

Navigational assistance service (NAS)

A navigational assistance service assists onboard navigational decision making and monitors its effects. This service is especially important in difficult navigational or meteorological circumstances or in the case of defects or deficiencies. NAS is given at the request of a vessel, provided it meets certain navigational equipment criteria or if the VTS considers it necessary for reasons of navigational safety.

One example where navigational assistance may be required is if the pilot onboard suddenly collapses, and the Master, who may not be familiar with the waterway that the vessel is transiting, needs assistance in conducting the vessel to safety. Another example is where a vessel appears to be navigating erratically or where navigational behaviour is not normal for the area in question.

It is important that if the VTS is authorised to issue instructions to vessels during NAS or indeed at any time, the instructions should be only result-orientated. This leaves details of execution, such as course steered or engine manoeuvres to be executed, to the Master or pilot onboard the vessel. Whatever the circumstances, the Master is still ultimately responsible for the vessel's safe navigation.

In order to achieve the objectives of providing NAS, good radio communications are necessary, the ship should be fitted with the appropriate navigational equipment and crewed with personnel capable of understanding and executing instructions from the VTS.

Traffic organisation service (TOS)

A traffic organisation service aims to prevent the development of dangerous maritime traffic incidents and to provide for the safe and efficient movement of vessels within the VTS area. This service concerns the operational management of traffic and the forward planning of vessel movements to prevent congestion and dangerous incidents and is particularly relevant in times of high traffic density or when the movement of special transports may affect the flow of other traffic.

The service may also include establishing and operating a system of traffic clearances or VTS sailing plans, or both, in relation to priority of movements, allocation of space, mandatory reporting of movements in the VTS area, routes to be followed, speed limits to be observed or other appropriate measures considered necessary by the VTS authority.

Training and certification of VTS personnel

All operational VTS personnel in VTS centres are professionally trained and qualified in a similar manner to navigation officers on vessels. The VTS Committee of the International

Chapter 10
Vessel Traffic Services

Association of Marine Aids to Navigation and Lighthouse Authorities (IALA) produced the IMO-approved international training and qualification standards. Many former seafarers re-train to become VTS operators in order to continue their careers ashore.

VTS and AIS

Automatic identification system (AIS) is a very high frequency (VHF) radio broadcasting system that transmits and receives data which can be displayed on an electronic chart, computer display or compatible radar.

AIS can be used as an aid to navigation, by providing location and additional information on buoys and lights. AIS should always be used wisely and in conjunction with all other aids, to promote safe navigation. In 2001, the IMO adopted Resolution A.917(22) – *Guidelines for the Onboard Operational Use of Shipborne Automatic Identification Systems (AIS)*. SOLAS chapter V Regulation 19 includes the carriage requirements for AIS.

Unfortunately, some OOW have come to over-rely on AIS, using it as a sole source of information, particularly in collision avoidance. Like any navigational aid, AIS does have limitations. Input errors by one vessel are reflected in the reception by other vessels and VTS centres. Sometimes AIS targets may not complement radar targets, and although many small vessels, such as tugs and yachts, do carry and use such systems, there are those that will not be fitted with it. The close proximity of bridges and the density of shore-side infrastructure in an area may also affect the AIS signal.

The OOW should also be aware that AIS fitted on ships as a mandatory carriage requirement might, under certain circumstances, be switched off. This might be the case where international agreements, rules or standards provide for the protection of navigational information. AIS is by no means perfect, and should be used as one part of a network of navigational aids fitted on the bridge, together with the keeping of a proper lookout.

Accidents and incidents in VTS areas

Unfortunately casualties still occur in spite of all the navigational aids carried on a vessel and the aids to navigation (AtoN) around coastlines and in the various waterways. Collisions and groundings probably account for the highest number. The causes are varied and include human error, fatigue, meteorological conditions, misinterpretation of data shown on navigational aids, lack of situational awareness and bad or improper communications, both internally on the vessel and externally, ship to shore.

All VTS authorities carry out risk assessments for their areas of responsibility, and update their safety management systems when necessary. Vessels should also carry out a form of risk assessment for their voyages as part of contingency planning, especially before entering or leaving ports.

There are many VTS related casualty reports available on the internet posted by the major marine accident investigation authorities. These reports provide an interesting insight into a particular incident, without pointing the finger of blame at any one person or authority.

The future of VTS

Advances in VTS equipment technology are being influenced by the need for a more comprehensive global traffic information system. This will lead to an increase in the volume of information being exchanged between the vessel and the individual VTS and between different VTS. This requirement for information exchange will lead to the formulation of international, as well as national, VTS networks.

Unlike the aviation industry, which has tightly-regulated aircraft and equipment specifications, the shipping industry, both ashore and afloat, has many difficulties to overcome before the global concept can become a reality. Traditionally, Masters are used to making their own decisions regarding the safety of their vessels. However, as the quality and accuracy of vessel tracking improves, the possibility to control traffic by means of instructions, rather than just information and advice, is likely to be used more widely.

The concept of eNavigation is to incorporate new technologies in a structured way and ensure that their use is compliant with the various navigational and communication technologies and services that are already available, providing an overarching, accurate, secure and cost-effective system, with the potential to support global coverage for ships of all sizes.

All commercial aircraft carry flight management computers (FMC), which hold all the data the Captain inputs with reference to the flight, from the departure gate of one airport to the arrival gate of another. This data includes full details of the flight plan, which is then filed and sent to all the authorised interested parties. A similar arrangement could be made for vessels, so that owners and authorised interested parties would be able to monitor a vessel's progress from one port to another.

However, as technology advances, so the lines of communication become more fragile and susceptible to unauthorised access. A holistic approach is necessary, which will require secure authentication at all times. The risk of network gridlock will also need to be mitigated.

The future goal, then, is to have a secure, interoperable global management system, for all authorised stakeholders during all phases of a vessel's voyage, which meets agreed levels of safety, provides for optimum economic operations, is environmentally sustainable and meets strict international and national security requirements.

This form of management would be known as sea traffic management (STM), which is the concept of sharing and using all information in order to improve the safe movement of all vessel traffic, thereby having a positive environmental impact as well as improving the efficiency of maritime transport in general. It will also have a positive effect on global security, including search and rescue operations.

Chapter 10
Vessel Traffic Services

VTS will continue to play an increasing part in improving safety to shipping worldwide, at the same time upholding its core mission to contribute to the safety of life at sea, the safety and efficiency of vessel traffic and the protection of the marine environment.

Rena (image: The New Zealand Herald/newspix.co.nz)

Chapter 11
Learning from accidents and near misses

By Captain Leslie R Morris

At one stage of my career I became a lecturer at a nautical college. This was when I discovered the benefit of collective knowledge. I learned from experienced lecturers that often the best way to teach was to ask the class questions. This was particularly true in classes of mates and Masters, as in those days the students already had substantial sea time and experience. They needed little encouragement to talk about their own experiences, so all that remained was for the lecturer to guide the discussion in order to satisfy the requirements of the course syllabus. In this way, individual knowledge was shared with others, including the lecturer. I would not be the first lecturer to admit that I received at least as much from mature students as I transmitted.

What has this got to do with anything?

For classroom, lecturer and students, substitute: ship, Master, crew. There is a lot of knowledge in each ship, and crew members have their own experiences to offer. Junior ranks may have less to offer than their seniors, but their knowledge will often be more up to date; encouraging them to participate benefits all.

That brings me to the purpose of this chapter: Learning from accidents and near misses. Understanding how incidents occur and how they can be prevented is an essential part of that learning. Let us now look at how this can be done, but before that it may be worth looking briefly at some IMO Codes and Conventions relevant to the subject.

The ISM Code

This says there must be procedures in safety management systems for reporting, investigating and analysing accidents and non-conformities. From this, the expectation is that safety and pollution prevention will be improved and measures introduced to prevent recurrence.

The ISM Code is now well established within the shipping industry. It is indeed the cornerstone of good safety management but it relies on the commitment, competence, attitudes and motivation of individuals at all levels for it to be effective.

Chapter 11
Navigation Accidents and their Causes

Casualty Investigation Code

This is the short name for the IMO Resolution MSC.255(84): The Code of The International Standards and Recommended Practices for a Safety Investigation into a Marine Casualty or a Marine Incident, which came into force in 2010.

The Casualty Investigation Code emphasises that marine safety investigations should not apportion blame or determine liability. This point is made at the outset, in Chapter 1 of the Code.

There is a tendency for the criminalisation of seafarers and The Nautical Institute and its members are increasingly concerned about this. However, the Casualty Investigation Code makes it very clear that the findings of marine accident investigation reports cannot be used as evidence in a court of law.

Sources of learning

Marine accident investigation reports

In accordance with the Casualty Investigation Code, most maritime flag states produce their own reports; others, especially the smaller ones, may call on other countries to assist. In any event, the reports in the public domain are many and varied. Most contain a brief overview of the incident and a list of recommendations. The full reports contain a great deal of data and can then be read as, and if, required. The main purpose of these reports is to establish what happened, and in so doing, to prevent it happening again.

The following list is a primary source of marine accident investigation reports. It is short but contains most of the marine accident reports that have shaped the international marine industry as it is today.

- Marine Accident Investigation Branch (MAIB), United Kingdom www.maib.gov.uk
- National Transportation Safety Board (NTSB), USA www.ntsb.gov
- Australian Transport Safety Bureau (ATSB) www.atsb.gov.au
- Transport Safety Board JTSB, Japan www.mlit.go.jp/jtsb/marrep.html
- Maritime Safety Administration, China www.en.msa.gov.cn
- Marine Department, Hong Kong, www.mardep.gov.hk

The websites of the six administrations, while each is different in its own way, provide a large database of marine accidents and near misses, and in the context of this chapter are useful sources of learning. Two maritime states will often report on the same incident, usually in accord with one another, but not always. A full list of the marine accident investigation organisations can be found at www.maiif.org

MARS

The Mariner's Alerting and Reporting Scheme (MARS), is a confidential resource compiled by The Nautical Institute to allow full reporting of accidents (and near misses)

Chapter 11
Learning from accidents and near misses

without fear of identification or litigation. It is a free service to the maritime industry and provides a valuable database on shipboard, inter-ship and other incidents. Members of The Nautical Institute and others will be familiar with its content. It is available on The Nautical Institute website www.nautinst.org or contact the editor at mars@nautinst.com

CHIRP

CHIRP is the UK's aviation and maritime confidential incident reporting programme which is a valuable resource for maritime safety worldwide. Like MARS it is a totally independent, confidential, but not anonymous, reporting system for all those employed in or associated with maritime industries. See www.chirp.co.uk

P&I clubs

All major P&I clubs have sophisticated websites and keep in touch regularly with their members. When a particular risk is identified an alert will be sent out. Some produce DVDs and other useful training aids.

Internet

Much information about marine incidents is available on the internet. Not all sources are reliable, but frequently have links to other, more reliable sites which may be helpful.

Others

There are many text books available. The Nautical Institute's *Seaways* magazine provides regular updates on best practice and navigational developments. The Marine Guidance Notes (MGN) and other documents available from the UK Maritime and Coastguard Agency (MCA) are accessible to all ships, not just UK flag.

Incidents

Voyage data recorders have helped improve the quality of evidence available to marine accident investigation. In the past, investigators had to rely heavily on witness evidence, much of which was unreliable. Keeping in mind the objectives of this publication – to encourage learning from accidents and near misses – a few examples are included below, with references and weblinks to encourage further investigation by readers. Some of the examples are relatively new, while others are not. Generally speaking, the consequences of marine navigation incidents can be divided into three sections:

- Near misses
- Groundings
- Collisions.

Brief examples of each of these have been extracted from their accident reports; the sources are shown. They are selected to illustrate some common problems that have arisen, but by no means all. That would be beyond the scope of this chapter.

Chapter 11
Navigation Accidents and their Causes

At this point, I am pleased to acknowledge the following marine accident investigation authorities; they have readily allowed me to refer to their reports, and in some cases to quote extracts. They are the MAIB UK, ATSB, Marine Department, Hong Kong (details as before), Transport Accident Investigation Commission, New Zealand (www.taic.org.nz) Maritime Survey, Gibraltar Maritime Administration (www.gibraltarship.com) and Accident Investigation Board, Norway (www.aibn.no/report).

Near misses

The adjective 'near' relating to miss is difficult to define, but we all know when we've had one! My own case is memorable – to me at least. I was senior OOW of a passenger liner in the Bay of Biscay on a clear dark night looking at a ship about two points on my port bow on a reciprocal course. It was showing a strong red sidelight and shaping to pass well clear down my port side – when it suddenly altered to port and started to show its green sidelight. My ship was doing 22 kt and I, and many others onboard, learned how much it could heel over in a full helm turn to starboard! It turned out the other ship had experienced a steering gear failure, according to an equally shaken German second officer who called me a few minutes later on VHF. I rarely took my eyes off an approaching ship ever again, regardless of CPA!

With the technology aboard modern ships, a near miss like that would be recorded and published. The following example is about one that was, and has become notorious as a result. See the MAIB report of the *Maersk Dover*, near miss with *Apollonia* and *Maersk Vancouver*, Dover Strait, October 2006.

Groundings

As with most shipping accidents, human error is often found to be the main cause, usually with good reason. However, such incidents are sometimes worthy of closer examination. For example, why did the human make the error?

Three groundings are good examples:

Case studies

Pasha Bulker, near Newcastle, NSW, Australia, June 2007 (Chapter 8, pages 74 and 79). The ship delayed acting on advice from authorities to move ahead of a storm. There were no casualties and no pollution.

Fedra, Gibraltar, October 2008. The ship was anchored to its port anchor for engine repairs but the Gibraltar Authorities were not informed at any time that the ship's engine was disabled. The weather deteriorated as forecast and the anchor started dragging. Although the starboard anchor was dropped about an hour later, the ship continued to drag.

There followed an extended period of confusion as tugs were called by variously the Authorities, the Master and the vessel's managers, but to no avail. *Fedra* continued to drag its anchors and grounded, broke in two and was wrecked. The crew were all saved, but the Gibraltarian and Spanish shorelines were extensively polluted.

Full City, Såstein, Norway, July 2009. The vessel grounded on a rocky coast. The hull was heavily damaged and there was a great deal of pollution. All the crew were safely evacuated.

This report is in itself a useful source of information of similar incidents, not just in the same location, but in other parts of the world, including the *Pasha Bulker* in Australia two years previously.

These three grounding casualties have common features; all three ships were at anchor on a lee shore, with severe weather forecast. All were in ballast, *Pasha Bulker* being especially light. The individual reports are very detailed and, not unexpectedly, raise a number of issues relating to the actions of the Masters and crew that contributed to the groundings and, in the case of *Fedra*, total loss of the ship. While mariners might be sympathetic to the contributory pressures brought to bear on their fellows by commercial aspects etc, we should also keep in mind that the basics of safety and good seamanship are paramount at all times. Anchoring on a lee shore is always potentially dangerous; if you have to, make sure your ship is ready in all respects to leave safely at short notice.

There are, of course, many other types of grounding. Probably the most common are those caused by navigation error. Examples include:

Case studies

Shen Neng 1, Queensland, April 2010. The investigation found that the grounding occurred because the chief officer did not alter course at an appropriate waypoint. His monitoring of the vessel's position was ineffective and his actions were affected by fatigue. The ship suffered severe bottom damage, with six tanks and the engine room breached. Pollution was relatively slight.

Rena, Astrolabe reef, New Zealand, October 2011. The final waypoint, at which *Rena* was to alter to port towards the pilot station, was 2 miles north of Astrolabe reef. The records show that the vessel was already south of the intended course at that time, and that subsequent small alterations to port made the situation worse. The ship was heading directly towards Astrolabe reef in the final minutes before it struck at 17 kt. Weather conditions were good at the time of grounding.

Rena eventually became a total loss, breaking up over the ensuing months. The subsequent accident investigation report concluded that the grounding was solely due to human error. The Master and second officer were sentenced to seven months in prison.

Chapter 11
Navigation Accidents and their Causes

Ovit, Varne Bank, Dover Strait, September 2013. (Chapter 4 and Chapter 9 pages 34 and 86) *Ovit* was aground for about three hours, there were no injuries and damage to the ship was superficial. There was no pollution. The ship's primary means of navigation was ECDIS and the MAIB concentrated largely on its installation, operation and aspects relating to training in its use. The main findings of the report included: the passage plan was unsafe as it passed directly over the Varne Bank; the OOW followed the track on the ECDIS display, but did not realise his ship was aground for about 19 minutes; the ECDIS safety settings were not appropriate to the local conditions and the audible alarm was disabled; although the lights of the cardinal buoys were all working, and were seen by the lookout, they were not reported, and that the OOW concentrated solely on the intended track on the ECDIS and ignored all other navigation aids in the area.

The *Ovit* full report is very comprehensive as it identifies many issues related to ECDIS that are now being addressed internationally within the maritime industry. The recommendations listed encompassed not just the ship and owners and the maker of the ECDIS equipment on board, but submissions and proposals to the IMO, Transport Malta, the UK's Maritime and Coastguard Agency, the International Chamber of Shipping (ICS) and the Oil Companies International Marine Forum (OCIMF).

These three groundings all differ in their causation, although they were all due to human error in the final analysis. However, other contributory factors were raised. For example, there is little doubt that the chief officer of *Shen Neng 1* was excessively fatigued and probably should not have been the OOW on the bridge at the time of grounding.

The *Ovit* grounding seems at first to have little or no contributory factors on the basis that the grounding would not have taken place had the vessel not had an ECDIS, but had used a paper chart and basic navigation skills. However, the fact that the listed recommendations of the report have been submitted to so many international bodies surely gives pause for thought.

Collisions

It has often been said that if Colregs were complied with at all times, then there would be no collisions. That may be so, but collisions continue to happen. These two examples are of collisions that took place nearly nine years apart. Similarities are obvious.

Case studies

Hyundai Dominion and *Sky Hope*, East China Sea, June 2004. The two vessels collided in good visibility. As the vessels approached, the OOW on *Sky Hope* incorrectly assumed *Hyundai Dominion* was overtaking his vessel. Action by both vessels was delayed by discussions on the VHF. Then the OOW on *Hyundai Dominion* used the free text facility on the AIS to request *Sky Hope* to keep out of the way. Action was taken too late and they collided. There were no injuries or pollution. Only *Sky Hope* suffered any significant damage. Each vessel was able to continue passage.

Clearly, several collision avoidance Rules were infringed in this collision and, to make matters worse, time was lost needlessly by talking on the VHF and texting on the AIS, instead of taking action as directed by the Rules.

CMA CGM Florida and *Chou Shan*, open sea, 140 miles east of Shanghai., March 2013. This was a more serious collision. Inappropriate use of VHF was again a major contributing cause of the collision, complicated by language misunderstandings. The VHF radio conversation onboard *CMA CGM Florida* was conducted in Mandarin by the Chinese supernumerary second officer, who was on a familiarisation voyage. *CMA CGM Florida*'s Filipino OOW did not understand Mandarin. After a great deal of confusion, both vessels altered course to port, which resulted in a continued risk of collision with each other. After further conversation on VHF both vessels altered course to starboard, collided and suffered heavy damage. Neither vessel had made any effort to slow down.

These two collisions clearly illustrate the dangers of inappropriate use of VHF communication in such a manner that several sections of the Colregs are ignored or infringed. What is worrying is that lessons are not being learned. These two collision cases were nearly nine years apart and there appears to be little or no reduction in the number of cases where inappropriate use of VHF is a contributing factor in the incident.

Rule infringements that commonly arise can be placed into four sectors, although, as even a brief review of accident investigation reports will reveal, it is rarely that only one rule is infringed in any single collision. The four sectors are:
- Inadequate lookout (Rule 5)
- Failure to proceed at a safe speed (Rule 6)
- Failure to assess risk of collision (Rule 7)
- Not taking early and positive action to avoid collision (Rule 8).

It may be interesting for readers to establish how many rule infringements they can find when they next read a marine accident report. A useful point of discussion for a safety meeting perhaps?

Finally...

Navigating officers have an array of navigation and collision avoidance aids at their disposal. Radar, AIS, VHF radio and ECDIS are compulsory in many ships. Let us not forget that the need to keep a lookout, proceed at a safe speed, assess risk of collision and take early and substantial action to avoid a collision are also legal requirements and failure to comply with them can increasingly result in criminal action.

A vast database of detailed information relating to ship collisions and groundings is now freely available to seafarers. Most, if not all safety information between ships and shore is transmitted via the internet. Information relating to significant casualties can therefore be sent to and from ships and downloaded for use. It is disappointing that some managements do not make better use of these resources.

Chapter 11
Navigation Accidents and their Causes

Learning from accidents and near misses is essential. It is important for ship's staff to have access to the relevant data; the Master and safety officer should then ensure that the information is assimilated properly aboard ship – as part of the ship's safety management system.

Further reading

IMO Resolution A.1075 (28): Guidelines to assist investigators in the implementation of the Casualty Investigation Code.

IMO MSC-MEPC.3/Circ.4/Rev.1: Revised harmonized reporting.

A N Cockcroft and JNF Lameijer. *A Guide to the Collision Avoidance Rules*, 7th edition, 2011

Dr Phil Anderson. *The Mariner's Role in Collecting Evidence*, 3rd edition, 2006

Steamship Mutual DVDs: *Collision Course; Groundings – Shallow Waters, Deep Trouble*, www.steamshipmutual.com

Chapter 12
Onboard training and mentoring

By Captain André L Le Goubin

I hope that by the time you have got to this last chapter you are not feeling too depressed and disillusioned having read so much about casualties and their causes. It must be remembered that the vast majority of voyages are undertaken without incident. Even so, accidents do occur and, if you remember back to this book's introduction, the frequency is not decreasing. This is even though technology is improving all the time and the modern ship's bridge is regularly fitted with one device after another to aid in accident avoidance.

In this chapter we will look at the onboard training and mentoring of seafarers and discuss if the quality of this training (or the lack of it) is a root cause of the recurrence of accidents. We will also discuss what can be done to ensure that Masters and watchkeeping officers get the maximum amount of training, particularly onboard training. This should be consistent with the requirements of safely operating on today's modern merchant navy vessels.

Before we go on, I will use some terms that may not be familiar so I will put them in a maritime context:

Mentor

The *Oxford English Dictionary* describes a mentor as 'an experienced and trusted adviser' and says the word originates from Mentor, adviser to the young Telemachus in Homer's *Odyssey*. In the context of this chapter, I simply define it as the possessor and distributor of experiential knowledge.

Mentoring

I define mentoring as the act of sharing knowledge without a designated reward and this is what differentiates mentoring from teaching and coaching. These two have tangible rewards.

Experiential knowledge

This is knowledge gained from professional on the job experiences which is then reflected upon. This knowledge can come from a wide variety of sources or experiences but, in my opinion, often has the most impact when it comes from an accident, incident or near miss. However, knowledge does need to be reflected upon before it can become experiential learning.

Chapter 12
Navigation Accidents and their Causes

Reflection

According to the Institute of Work Based Learning, reflection is a "thoughtful" consideration of your experiences, which leads you to decide what they mean to you. Over the years, reflection has become a very useful tool for me, particularly to review the actions I have taken and to help me become comfortable with the decisions I have made. I use reflection to review those decisions and decide my course of action when faced with similar situations in the future. In my book *Mentoring at Sea – The 10 Minute Challenge,* published by The Nautical Institute, I have a chapter devoted to reflection for seafarers and I hope that you will take time to read it.

Throughout these chapters many causes of collisions and strandings have been discussed, along with ways to prevent these catastrophic events. For that is what they are – catastrophic. As a marine consultant, I can remember walking up and down the deck of a fully loaded tanker which had grounded. I was just listening to the Master talk and trying to be a support to him, as he came to terms with the situation. It was emotional enough for me, let alone that poor man as he wondered how he and his family were going to manage in the future.

As seafarers we need to try to prevent these situations. In my opinion the best way to do this is by increasing the amount of onboard training and mentoring that the Masters and navigating officers undertake and receive.

Most seafarers involved in a navigational accident hold a valid STCW certificate and, to gain this certificate, will have gone through a substantial amount of training at a maritime training establishment before the incident. I am often asked if it is this training that is letting the maritime industry down. There are indications, currently being discussed in the maritime community, that the standard of teaching in some countries is falling below acceptable levels. This is something that the responsible governmental organisations need to address worldwide. In my experience the educational standard of candidates leaving a maritime training establishment is much the same as it always has been, although the method of teaching has changed substantially.

Another major change has been in educational facilities, primarily the extensive use of simulation to replicate onboard vessel operations. Simulators are now highly sophisticated training tools, but do these simulators really represent and reflect the modern ship's navigation bridge? They certainly look and feel similar but they may not recreate real life.

For example, in a simulator navigators will be concentrating 100% on the task in hand – be it ship handling, navigation, collision avoidance, etc. Phones do not ring, there is not a constant stream of emails to be dealt with, many marked urgent and there are no domestic issues. Importantly, at the back of their minds is the realisation that a fundamental mistake in the simulator may hurt the ego, but it will not cause death, injury or damage to the environment. The other major difference between simulators and real life is that simulators can be rewound and replayed, real life actions cannot!

Chapter 12
Onboard training and mentoring

Notwithstanding that, simulators are great tools, and must be used extensively, at every opportunity, to train seafarers, but to supplement, rather than replace, onboard training.

It is not educational training that has led to many of the navigational accidents we have discussed so far, but the lack of, or limited amount of, onboard training. The majority of educational establishments do a good job of preparing their candidates for their various levels of STCW Certificates of Competency but this academic training must be supported by a substantial amount of both structured and unstructured training onboard ships. In simple words, the opportunity to put theory into practice and gain experiential knowledge at every level, to ensure safe navigation and avoid the accidents and incidents we have discussed.

According to current belief, approximately three-quarters of our knowledge comes from experience. It is this knowledge gained from so many circumstances that, when reflected upon, can be turned into experiential knowledge which can then be used practically.

For example, if you have read your way through this book up to this point you will have gained a substantial amount of knowledge about the causation of maritime accidents. Take a few minutes out of your busy schedules to quietly reflect on what you have learned from this book and how you could use that knowledge. By doing this you will have turned the knowledge you have gained from your reading into experiential knowledge which you can use or pass on.

One of my greatest concerns is the practical training of cadets, as most of those I sail with have to complete only 12 months of sea time after an extensive period of three or four years at a maritime training facility. During this sea time they will undergo onboard training and need to gain sufficient experiential knowledge to become competent (and confident) watchkeeping officers.

This does not take into account the time it may take young seafarers to settle down onboard and integrate into an environment that is often multicultural and away from home, which we all had to get used to. All this is often without access to social media, which is a big issue these days and which needs to be addressed. This settling down period can sometimes be extensive and rapidly consumes the limited time for effective onboard training.

In a generation, the sea time requirements for cadets have halved or, depending on the country, are even one-third of what they used to be. I often wonder if I was competent to hold a navigational watch at the end of my first 12 months sea time. One company I am familiar with has addressed this concern and requires its newly qualified deck officers to complete an additional 12 months sea time as fourth officer. A practice I totally support.

During their onboard training, cadets must be exposed to as many navigational conditions during the time available. Where possible they must be allowed to conduct the navigation of the vessel, under supervision, particularly toward the end of their training. To do this successfully, they should spend most of their time with the senior navigation officer. These senior officers should be confident enough in their own experience to allow cadets to experience navigating the vessel.

Chapter 12
Navigation Accidents and their Causes

Another area of major concern is with the chief officers' onboard training, especially those aspiring to command. For example, so many companies now mandate through their onboard ISM system or SMS that the chief officer is to be forward for mooring stations and anchoring the vessel. A few years ago I was discussing this with one Master and he showed me a letter instructing him in this procedure. This stated that in that company chief officers were not allowed on the bridge for anchoring or mooring stations until they were formally training for command, when they would be taken out of the watchkeeping roster for three months.

How can chief officers gain experiential knowledge of handling the ship under a Master's supervision when they are stationed on the fo'c'sle? Of course they can't. This is a lost opportunity, but it can be overcome with the cooperation of shipboard and shore management. In this case, a change needs to be made to the vessel's SMS to allow other officers to be stationed forward and release the chief officer to understudy the Master on the bridge. Other officers must be trained to go forward and, in my experience, this is a fairly easy process.

North P & I Club statistics show an increase in anchoring accidents occurring worldwide (*Signals* 2010). I have investigated incidents that occurred when new Masters were trying to anchor their vessels in less than ideal environmental conditions and I have found that lack of experience was a root cause of those accidents.

But what about other officers? Every officer onboard modern merchant vessels should be training for their next promotion and should be training their successors. To succeed, this methodology of onboard training has to be started and supported by the Master and work its way down from there. This is not enough – it is imperative that this policy is fully supported by shore management. Before promotion, officers' practical skills should be assessed to ensure they have reached the necessary level to be competent in the next rank.

I know how busy Masters of today's merchant vessels are, but I urge them to spend just 10 minutes a day with each of their navigating officers. That is the time it takes to smoke a cigarette or drink a cup of coffee. In this time pass on a piece of your experiential knowledge to them or (gently) address a concern you may have.

For example, in my night orders I always put: Call me if in any doubt; but does the officer know the definition of being in doubt? After I have written the orders I go through them one by one with the duty officer and I often ask them how would they know they were in doubt? We discuss this in an informal fashion but when I leave the bridge I am reassured that the officer would call me if I am needed. That is one of the fundamental benefits of mentoring, and although there are no designated rewards for mentoring, over time you will gain the benefits of your efforts – if only by sleeping more soundly.

As the Master sets this example, so all the other officers should be encouraged to do the same thing with their immediate juniors; chief officer to second officer; second to third and all should be mentors to the cadets. What elements of your experiential knowledge to pass on is, to a great extent, personal preference. To help with this I would

Chapter 12
Onboard training and mentoring

like everyone who is engaged in mentoring to take my 10 minute challenge described in my mentoring book.

During that 10 minutes of reflection decide what your greatest concerns or needs are, or the anxieties of the seafarer to be mentored. Then address those concerns in a positive manner, perhaps using anecdotal evidence to bring these to life. You could use casualty examples, such as those in this book, or case studies you are familiar with or have found for this purpose. The internet is a huge resource for this, if you have access. If you do not, maybe this is a way of helping to persuade your company to install it.

Over time this style of experiential knowledge transfer will often develop a bond of confidence between the navigators and the Master and you may be surprised by what gets discussed. Near misses, close quarters situations, even casualties, and, of course, the value is in the knowledge that is to be gained from one another's experiences.

Mentoring need not take extra time if it is undertaken while on the bridge during the course of your normal duties. I remember one occasion when I was third mate onboard one of the (then) world's largest ro-ro/container ships. It was a Sunday morning and we were transiting a fairly busy Dover Strait north bound towards the port of Antwerp.

It was usual on these occasions for the Master to take the con and for me to assist him but on this occasion he did not. He allowed me to continue navigating the vessel. It is almost 30 years ago but I still remember vividly how exciting and fulfilling that was for me. I knew the Master was keeping a good eye on what I was doing but the effect, to this day, is tremendous.

We have talked a little about onboard training of officers but what about when it is the Master that needs the training? In my book I have defined four critical areas where, if the Master is well trained, many accidents can be avoided. They are:

- Anchoring the vessel
- Manoeuvring to embark a pilot
- Manoeuvring the vessel in a channel or fairway
- Handling the vessel in periods of heavy weather.

These four areas I have defined come from a substantial amount of research, including the review of many accident reports. On the surface they may appear somewhat simplistic, but if Masters can safely undertake these four tasks then they will be closer to operating the ship safely.

The best time for Masters to gain this onboard training is before they are promoted but, if Masters identify that they need training (or need further training), much can be achieved by simulation, with the exception of handling the vessel in heavy weather. However, the need must be identified, preferably before an accident, and the Master's company needs to be sympathetic to, and accommodating of, these needs.

For example, with the exception of real-time handling of a vessel under supervision, a manned model course is by far the best way to gain realistic experiential knowledge in ship handling. These courses can be expensive but, in terms of value for money, they are

Chapter 12
Navigation Accidents and their Causes

well worth it, especially if they prevent costly accidents. Of course, in the short term, we will never know if this is the case. However, I believe that, in the long term, companies that actively engage in hands-on training of their officers will see the benefits through an auditable reduction in accidents and insurance claims.

So, how to proceed? Mentoring as an informal method of transferring experiential knowledge is as old as seafaring itself, yet it is still the most effective way of undertaking and achieving training onboard today's modern merchant navy vessels. Mentoring is most efficient when it is incorporated into the daily routine of ships when everyone onboard is engaged with it as a usual, rather than extraordinary, event. In this way, mentoring should never take more than that daily 10 minutes extra of your precious time to undertake as, most of the time, the task has to be undertaken anyway.

Each officer, as soon as they are settled in their rank should be encouraged to start training for the next position and equally, it should be firmly established that each officer that is training for the next level should also be training their successor. This style of training will help to ensure that experiential knowledge is transferred from one seafarer to another to break down the barriers that often prevent a more effective formal onboard system of training.

Having read this chapter, I hope that you will now reread and reflect on some of the accidents and incidents that have been discussed in this book. Consider whether a lack of experiential knowledge that could have been gained from onboard training and mentoring may have been a root cause of that accident. If you consider that it may have been, and you are in a position to be a mentor, use your position as a way of sharing experiential knowledge. Perhaps, by doing so, you will prevent a similar accident taking place in the future.

If mentoring is established (or re-established) on ships as a routine daily activity then many of the types of accidents and incidents we have been discussing throughout this book can be avoided. It is my belief that we are all members of the maritime community, whatever our position. As we learn and advance, so we have a duty to pass on our knowledge to those following on after us in that community. By being a mentor, and passing on your experiential knowledge, you will be fulfilling that traditional maritime obligation.

Index

A

Accident reports 3–4, 102, 103–4, 113
   see also Case studies

Accidents see Navigation accidents

AIS (automatic identification system) 11, 13, 27
   collision avoidance 42–43, 47, 97, 106
   familiarisation 83
   limitations of 97
   VTS and 97

Alandsfarjan 88

Alarms
   mis-used 4
   passage plans and 12
   turning off 85, 86–8, 106

Alcohol, affecting performance 3

Altering course 43, 44–5

Anchoring 71–81
   anchor watch 79–80, 81
   anchoring prohibited 77, 78
   case studies 74, 76, 78, 104–5
   collision with other vessels and fixed or floating objects 77–78
   contingency planning 80
   currents and tidal streams 74–5, 76, 77
   damage to anchor, cable and mooring equipment 78
   dragging of anchor 73, 74, 76, 77, 78, 79, 80, 81, 104
   excessive yawing 76
   exposed location 76–7
   fouling of anchor and cable 78, 79
   grounding or impact with seabed 74, 76, 77, 79, 104–5
   holding power of anchor and scope of cable 73–4, 75
   Masters and 72, 74, 75, 78, 79, 80, 81
   planning for anchoring 72–3
   position monitoring 79–80
   readiness of main engine 80–81
   risks of 71–2, 73, 77–9, 112
   security 80
   submarine cables and pipelines 79
   water depth and 77, 78
   weather and 79, 80, 81
   wind and 75, 76, 77

Apollonia 104

ARPA (automatic radar plotting aid) 4, 13
   collision avoidance 40, 42
   positioning and 27, 29, 34

ATC (air traffic control) 92

Atlantic Blue 30–1

AtoN (aids to navigation) 32, 97, 106

ATSB (Australian Transport Safety Bureau) 30–31, 58, 59, 74, 79, 84, 102, 105

AUSREP (Australian Ship Reporting Area) 94

Autopilot
   collision avoidance and 39, 41
   over-reliance on 88, 90
   positioning and 33

B

BEAmer (French Marine Accident Investigation Office) 3

Beidou 15, 26

Index

BELTREP (Great Belt Ship Reporting System) 94

Bridges
    bridge design 4
    one-man bridge 5, 10

BRM (bridge resource management) 17–24
    briefings and debriefings 20, 21, 22, 23
    case study 21
    clarity of purpose 19
    common ownership of the plan 21–2
    competence and confidence 20–1
    concise communication 19–20
    conduct and proper use of checklists 18, 20, 21
    consistency 19
    credit and congratulation 22
    future trends 23–4
    navigational audits 23
    passage planning and 23, 24
    training 23, 95
    working as a team 19, 20–1, 22, 24, 50–1
    working language 19–20, 51–2

C

Capella Voyager 59, 63, 66

Cardiff Research Programme 2–3, 5, 6

Carnival UK 23

CASCADe project (Model-based Co-operative and Adaptive Ship-based Context Aware Design, 2012-2015) 3, 5

Case studies iv-vii
    anchoring 74, 76, 78, 104–5
    BRM and 21
    collisions 10, 84, 89, 104, 106, 107
    electronics 84–7, 89–90

groundings 12, 21, 22, 25, 30–1, 33, 34–5, 58–9, 74, 76, 78, 84, 85, 86–7, 88, 90, 104–5, 106
    passage planning 10, 12
    positioning 25, 30–1, 33, 34–35, 105–6

Casualty Investigation Code 102

Celestial navigation 26, 27, 31–2

Checklists
    BRM and 18, 20, 21
    pilotage 54

CHIRP (Confidential Hazardous Incident Reporting Programme) 22, 103

Chou Shan 107

CMA CGM Florida 107

Collision avoidance 39–48
    action to avoid collision 43–4, 46–7, 89, 108
    AIS, use of 42–3, 47, 97, 106–7
    in anchorages and port approaches 46
    ARPA, use of 40, 42
    autopilot and 39, 41
    future technology 47–8
    know your duty as watchkeeper 40
    know your ship 39
    knowledge of basic control measures 39
    lookout, keeping a proper 40–1, 107
    managing 44–5, 107
    radar, use of 41, 42, 43
    risk of collision 41–2, 46, 107
    route direction 48
    safe speed 44, 107
    stand-on vessel, action by 45–6
    traffic separation schemes 42, 46, 94
    use of VHF for 46–7, 106–7

Collisions
    accident reports 3–4, 102, 103–4, 113
    anchoring and 77–8
    case studies iv-vii, 10, 84, 89, 104, 106, 107

Index

electronics and 84, 89
learning from 106–7
passage plan failures 9, 16
pilotage and 51
in VTS areas 97

Colregs (International Regulations for Preventing Collisions at Sea) 39
action by stand-on vessel – Rule 17 45–6
action to avoid collision – Rule 8 43–4, 46–7
keeping a lookout – Rule 5 40–1
managing collision avoidance – Rule 8(d) 44–5
risk of collision – Rule 7 41–2
rule infringements 106–7
safe speed – Rule 6 44, 107

Communications
BRM and 19–20
failures of 3, 5–6
between pilots and crew 51–2
use of VHF for collision avoidance 46–7, 106–7
between watch members 14
see also VTS

Compasses 26, 33

Costa Concordia 12, 19, 20, 35

CPA (closest point of approach) 12, 41–2, 104

Crew 1–7

D

Desh Rakshak 59, 63

Dover Strait 13, 42, 85, 86, 93, 94, 104, 113

DP (dynamic positioning) 27, 32

DR (dead reckoning) 27–8

E

Eastern Honor 59

ECDIS (electronic chart display and information system)
advances in 54
checking settings 35, 84–7, 106
familiarisation 83
GNSS and 27
misuse 4, 12, 106
passage planning 10, 12, 13, 14
positioning and 35, 106

Echosounder 32, 33

ECS (electronic chart system) 27

Electronics 83–9
case studies 84, 85, 86, 87, 89, 90
checking settings 84–7
familiarisation 83
future technology 15–16, 47–48, 54–5, 98–99
jamming 15, 26, 38, 90
mental stimulation and 89
paying attention to 88
reliance and over reliance on 15, 38, 89, 90, 97, 106
see also specific types of equipment

ELoran 32, 33, 90

ENavigation 23, 32–33, 98

ENC (electronic navigation chart) 35, 36, 86

Energy-efficiency 54–5

Engines 54–5, 80–1

EP (estimated position) 27–8

Ever Decent 10

Experiential knowledge 109, 111, 112, 113, 114

F

Fatigue 1–7
    accident reports 3–4
    anticipating 11
    Cardiff Research Programme 2–3, 5, 6
    fatigue process 1
    managing 7
    manning levels, related to 4–5, 6, 7, 11
    signs and symptoms of 1–2, 3
    solutions to the problem of fatigue 6–7
    work-related fatigue accidents 1

Fedra 104–5

Ficus 25

Flag states 4, 6, 19, 102

FMC (flight management computers) 98

Froude depth number 62–3

Full City 105

Furness Melbourne 84

G

Galileo 15, 26

Glonass 15, 26, 32, 38

GMDSS (Global Maritime Distress and Safety System) 27, 40–1

GNSS (global navigation satellite system)
        failure of 15, 90
        familiarisation 83
        passage planning 10, 15
        positioning systems and 26–7, 33, 38

Godafoss 87–8

GPS (global positioning system)
        anchoring and 79, 80
        failure of 15, 89–90
        positioning with 26–7, 30–1, 32, 33–4, 38
        real-time-kinematic GPS 67

Groundings
        accident reports 3, 102, 103–4, 113
        anchoring 74, 76, 77, 79, 104-5
        BRM and 21, 22
        case studies iv-vii, 12, 21, 22, 25, 30–1, 33, 34–5, 58–9, 74, 76, 78, 84, 85, 86–7, 88, 90, 104–5, 106
        ECDIS and 85, 86–7, 106
        learning from 104–6
        passage plan failures 9, 12, 16
        positioning 25, 30-1, 33, 34–5
        in VTS areas 97

H

Health and Safety Executive 1

Heave, UKC and 63, 65–6

Heel, UKC and 57, 58, 64–5, 67, 69

Herald of Free Enterprise 1

Hilda 83

HO (hydrographic offices) 11, 77

Hyundai Dominion 106–7

I

IALA (International Association of Marine Aids to Navigation Lighthouse Authorities) 32, 97

ICS (International Chamber of Shipping) 106

IMO (International Maritime Organization)
        AIS and 97
        alarms and 12
        Casualty Investigation Code 102
        electronics 83

eNavigation 32
ISM Code 12, 15, 19, 101
pilot boarding and landing 50
radar performance standards 29
VTS and 91–2, 93–4, 96–7

IMPA (International Maritime Pilots' Association) 50

INS (information service), VTS 95–6

Institute of Work Based Learning 110

IRNSS (Indian Regional Satellite System) 26

ISM (International Safety Management) Code 12, 15, 19, 101, 112

ISPS (International Ship and Port Facility Security) Code 80

J

Jody F Millennium 58, 64–5

K

K-Wave 22

L

Lookouts
 collision avoidance 40–1, 107
 including in bridge team 13

LOP (line of position) 28, 29, 32

Loran 32, 90

LPS (local port service) 92

LRIT (long-range identification and tracking) 27

LT Cortesia 85, 86

M

Maersk Dover 104

Maersk Kendal 21

Maersk Vancouver 104

MAIB (Marine Accident Investigation Branch) 3, 21, 22, 34, 74, 76, 85, 86, 102, 106

Manning, fatigue and 1–7, 11

Marinecasualty.com 3

MARS (Mariners' Alerting and Reporting Scheme) 22, 102–3

Masters
 anchoring and 72, 74, 75, 78, 79, 80, 81
 BRM and 17–18, 20
 crew fatigue and 7
 mentoring junior officers 112, 113
 pilots and 5–6, 49, 50–1, 54, 59, 95
 presence on bridge 10–11, 12, 40
 training of 109, 110, 112, 113–4
 VTS and 95

MCA (Maritime and Coastguard Agency) 103, 106

Mental workload, crew members 4, 16

Mentoring
 BRM 23–4
 definition 109
 onboard training and mentoring 109–114
 passage planning 16

MGN (Marine Guidance Notes) 103

MPX (Master/pilot exchange) 50–1

MSP (marine spatial planning) 15

MSRS (mandatory ship reporting systems) 94

Index

N

NAS (navigational assistance service), VTS 96

The Nautical Institute
  ECDIS familiarisation 83
  electronics familiarisation 83
  MARS 22, 102–3

Navigation accidents
  accident reports 3–4, 102, 103–4, 113
  contributory factors 3–4
  learning from 101–8
  passage planning failures 9, 16, 34–5
  in VTS areas 97–8
  see also Case studies; Collisions; Groundings; Near misses

Navigational audits 23

Navigational equipment
  deficiencies in 4, 93, 96
  see also Electronics

Near misses 14, 22, 89, 104
  learning from 101–8

Norwegian Dream 10–11

NTSB (National Transportation Safety Board) 102

O

OCIMF (Oil Companies International Marine Forum) 106

Office of Rail Regulation 1

Oliva 33

OOW (officer of the watch)
  AIS and 97
  anchoring and 74, 79, 80, 81
  BRM and 17–18, 21
  doubling the navigation watch 14
  electronics and 83–90
  executing passage plans 11–12, 13–14
  positioning and 25, 30–1, 33, 105–6
  see also Watchkeeping

Ovit 34–5, 86, 106

P

Parallel indexing 29, 30–1, 34

Pasha Bulker 74, 79, 104

Passage planning 9–16
  BRM and 23, 24
  case studies 10, 12
  common faults in planning 12–13
  doubling the navigation watch 14
  executing and monitoring the plan 11–12, 13–14
  future developments 15–16
  identification of risk 10–11, 14, 15, 97
  importance of 9
  information sources 11
  minimising risk 14
  owners' role 15
  planning failures 9, 16, 34–5
  safe water 12, 15

P&I clubs 103, 112

Pilots and pilotage 49–55
  boarding and landing of pilots 49–50, 91
  BRM and 17–18, 23
  future trends 54–5
  Master and 5–6, 49–54, 59, 95
  passage planning and 9, 15
  positioning and 30–1
  record-keeping and checklists 54
  remote pilotage or shore assistance 53
  status and role of 49
  teamwork and culture 51
  tugs and towage 52–3
  voluntary and compulsory pilotage 53–4

Index

Pitch
  anchoring and 76
  UKC and 59, 65–6

Positioning 25–38
  active navigation 34–5
  case studies 25, 30–1, 33, 34–35, 105–6
  celestial navigation 26, 27, 31–2
  competent positioning, 10 steps for 37–8
  DR (dead reckoning) 27–8
  eNavigation 32–3
  EP (estimated position) 27–8
  following a planned track 33
  heading systems 26
  other systems 32
  parallel indexing 29, 30–1, 34
  passive navigation 33–4
  positioning systems 26–7, 38
  radar fixing 29–31, 84
  safe water 35–7
  situational awareness 25, 33–5, 81
  terrestrial fixing 28
  see also Anchoring

Pride of Canterbury 34, 85

Princess Cruises 23

Q

Queen Elizabeth 2 58

QZSS (Quasi-Zenith Satellite System) 26

R

RACON (RAdar beaCON) 29, 33

Radar
  advances in 54
  checking settings 84–7
  collision avoidance 41, 42, 43
  port control 92
  position fixing 29–31, 84

REEFREP (Great Barrier Reef and Torres Strait Ship Reporting System) 94

Rena 105

Riga II 84

Risk
  of anchoring 71–81, 112
  of collision 41–2, 46, 107
  identifying in passage planning 10–11, 14, 15, 97
  VTS risk assessments 97

River Embley 58

Roll, UKC and 58, 59, 63–5

Ropax 1 74, 78

Royal Majesty 90

S

Safe speed 44, 107

Safe water
  passage planning 12, 15
  positioning 35–7

Safety contour 34, 85–6

Safety depth 34, 85–6

Sailing Directions 11, 13, 72, 73

Sea Diamond 35

Sea Empress 58

Sea time 101, 111

Shen Neng 1 105–6

Shipowners
  passage planning and 15, 22
  route direction and 48

Singapore Strait 13, 14, 21

Sky Hope 106–7

Sleipner 84

SMCP (Standard Marine Communication Phrases) 93

SMS (safety management system) 15, 19, 54, 72, 80, 112

SOLAS (Safety of Life at Sea Convention)
    AIS and 97
    MSRS and 94
    passage plans 9
    pilot boarding and landing 50
    VTS and 92

Squat
    formula to predict squat 62–3
    passage planning and 10
    safe water and 37
    UKC and 57, 58, 59–63, 66

Standing Orders 19, 77, 80

STCW (Standards for Training, Certification and Watchkeeping for Seafarers)
    BRM 95
    Certificates of Competency 110, 111
    passage planning 16
    VTS 93

Stena Alegra 76, 77

STM (sea traffic management) 98

SWL (safe working load) 53

T

Tarnfjord 89

TCPA (time to closest point of approach) 42

Titanic 92

TOS (traffic organisation service), VTS 96

Training
    bridge teams 13–14, 95
    BRM 23, 95
    experiential knowledge 109, 111, 112, 113–14
    learning from accidents and near misses 103–08
    onboard training and mentoring 109–14
    positioning systems 38
    sea time 101, 111
    VTS 96–7

Trans Agila 87

TSBC (Transportation Safety Board of Canada) 3

TSS (traffic separation schemes) 42, 46, 94

Tugs and towage 52–3

U

UKC (under-keel clearance) 12, 57–69
    anchoring and 77
    ECDIS settings and 85–7, 89
    grounding case studies 58–59
    heave and pitch 59, 65–6
    heel and 57, 58, 64–5, 67, 69
    passage planning and 9, 10, 37
    pilotage and 51–2
    roll and 58, 59, 63–5
    safety depth and safety contour 34, 86–7
    squat and 57, 58, 59–63, 66
    static UKC 57, 58–9, 63, 66–7, 69
    UKC management in practice 66–68
    UKC software 67–68
    unusual conditions 68–9

V

VDR (voyage data recorder) 27, 54, 103–4

VHF radio
    familiarisation 83
    inappropriate use of 106–7
    use of VHF for collision avoidance

46–7, 106–7
see also AIS

VTIS (vessel traffic information system), Singapore 21

VTS (vessel traffic services) 91–99
    AIS and 97
    anchoring and 80, 81
    BRM and 23
    creation of 91–2
    definition of 92–3
    future of 98–9
    limitations of 93
    MSRS 94
    passage planning and 14, 15
    remote pilotage 53
    role and responsibility of 94–5
    services provided by 95–6
    training and certification 96–7
    types of 93–4

W

Watchkeeping
    action to avoid collision 43–4
    anchor watch 79–80, 81
    ARPA and AIS, use of in collision avoidance 40, 41, 42–3
    keep a look-out 40–1
    know your duty 40
    know your ship 39
    managing collision avoidance 44–5, 106–7
    risk of collision 41–2
    safe speed 44, 107
    see also OOW (officer of the watch)

Wave buoys 59, 60

Waypoints 12–13, 26, 30, 105

Weather
    anchoring and 79, 80, 81
    contributory factor in accidents 4, 105

Weather reports 20–1

Wellamo 89

Wellpark 58

Willy 76

Windage area 52, 74, 75

Windlasses 78, 79

Work-related fatigue accidents 1

Working language 19–20, 51–2

Y

Young Lady 78, 79

Contributors

Paul H Allen MA BSc

Paul has been involved in seafarers research, mainly on fatigue, since joining the Centre for Occupational and Health Psychology at Cardiff University in 2002. He is also a professional film maker and in 2011 produced a 30-minute film on fatigue as part of an Economic and Social Research Council knowledge exchange grant. He is exploring ways of using film to conduct and disseminate research.

Captain Nadeem Anwar FNI MSc BSc PGCEL CertEd ACII

Nadeem works for Southampton Solent University in the Petrochemical Section, having had more than 14 years' experience on deep sea vessels, including in command of very large petroleum and gas tankers. He has produced a wide range of training materials on shiphandling and emergency response and written books on ballast water management, navigation and passage planning. His experience also includes complex marine projects and marine consultancy.

Captain Nicholas Cooper FNI MNM

Nicholas is currently a consultant with Cwaves, drawing on 40 years of seagoing experience of which 25 were in command on container ships and bulk carriers. He also has wide experience of ship and cargo superintendency and surveying and has implemented and audited quality and safety management systems. He is a Past President of The Nautical Institute.

Dr Tim Gourlay BSc BApplied Maths (Hons) PhD MPIANC AMRINA AMSNAME

Tim is a Senior Research Fellow at the Centre for Marine Science and Technology at Curtin University in Australia, working in shallow-water ship hydrodynamics. His principal research area is calm-water sinkage and trim. He has led an international benchmarking study into ship wave-induced motion codes, assisted with accident investigations on groundings due to squat and wave-induced motions and helped develop UKC guidelines for ports in Australia and internationally.

Contributors

Captain Robert Hone BA (Hons) MNI FHEA

Robert is currently a lecturer in Navigation and Maritime Science at Plymouth University, specialising in the human element, leadership and management. He is researching the wellbeing and happiness of seafarers. His seagoing career began in 1977 with Cunard Steam Ship and he gained his Master's certificate in 1990 with Curnow Shipping. He sailed as chief officer on the *QM2* and as Staff Captain on the *QE2*.

Captain Terry Hughes FRIN FNI

Terry established International Maritime Consultancy in 1996 and provides consultancy on training and operations for Vessel Traffic Services (VTS) and VTS incident investigation. After 22 years at sea, ending as Master with Bibby Line, he lectured for the Commonwealth Secretariat and at Warsash, specialising in VTS. He has been instrumental in obtaining professional and internationally recognised qualifications for VTS operators and is an accredited VTS expert with IALA's World Wide Academy.

Captain André L Le Goubin MNM MA FNI

André is currently employed as a Mooring Master in the lightering trade, undertaking ship-to-ship transfers in the Gulf of Mexico. He has recently started his own company, DNA Marine, where he specialises in providing expert witness services. His experience includes command of high-speed ferries (hydrofoils, mono- and multi-hull ro-ro/passenger vessels), working as a pilot for the London Pilotage Service and acting as a marine consultant.

Captain Leslie R Morris BSc (Hons) FNI MRINA

Leslie founded Con-Mar International in 1999 and retired recently as a consultant and expert witness in marine operations, specialising in towage, salvage and wreck removal. He gained his Master's Foreign-Going Certificate in 1973 and came ashore to lecture at Warsash in 1975, specialising in Control Engineering and Radar Simulation. While at Staveley Electrotechnic Services as Nautical Adviser and Master he was involved in ARPA development and geotechnic surveys.

Contributors
Navigation Accidents and their Causes

David A Pockett Master Mariner BSc FNI

David's prime professional interest is marine casualty investigations and management. He is on the panel of Special Casualty Representatives at Lloyd's. He had a seafaring career in general cargo ships and tankers in all ranks up to and including Master. He was a founding director of London Offshore Consultants, where he later became CEO. He has attended and investigated numerous casualties worldwide as consultant to owners, charterers, insurers, government bodies and salvors.

Professor Thomas Porathe PhD MNI

Thomas is currently professor of Interaction Design at the Norwegian University of Science and Technology in Trondheim, where he teaches Information Visualisation and Human Centred Design. He previously taught maritime human factors at Chalmers University of Technology, Sweden. He has been involved in human-machine interaction and information systems in eNavigation research projects including EfficienSea, MONALISA, ACCSEAS and the unmanned ship project MUNIN. He is also active in the IALA eNavigation Committee.

Professor Andrew P Smith BSc PhD FBPsS CPsychol FRSM

Andy is Professor of Psychology and Director, Centre for Occupational and Health Psychology, Cardiff University. He conducts research on stress at work, the working environment, transport fatigue, nutrition and behaviour, the psychology of acute and chronic illness, and wellbeing. He has held positions at Oxford University, the MRC Perceptual & Cognitive Performance Unit at the University of Sussex, and the University of Bristol.

John Hamilton Third Master Mariner Extra Master RSA (SM)

John is a senior consultant with the consultancy Brookes Bell, specialising in navigation, seamanship and collision analysis, cargo stowage and container securing system failures. He has extensive experience as an expert witness and over 15 years' seagoing experience sailing on general and refrigerated cargo vessels, bulk carriers, passenger liners and offshore supply vessels, serving in ranks up to and including Master.

Contributors

Emma J K Wadsworth BSc PhD

Emma is a Research Fellow at the Cardiff Work Environment Research Centre, Cardiff University, UK. Her research has focused on various aspects of the work environment, including work and wellbeing and workplace health and safety and its management and regulation.

Captain Paul Whyte MBE AFNI

Paul is a Master Mariner and consultant at London Offshore Consultants. He has provided expert advice for accident investigation, speed and performance, salvage dangers, grounding, fixed object and ice damage claims. He has 37 years' sea-going experience working on a broad range of commercial and primarily military vessels, and 12 years command experience on a variety of naval auxiliaries undertaking global operations.

Captain Richard Wild MSc FRIN MNI

Richard is Pilotage Manager at a major UK port, where he previously served as Senior and Examining pilot. He is a member of the Navigation Safety Committee and has participated in feasibility studies into terminal requirements and dynamic berth exchange procedures for ultra-large container ships. His seagoing career was mainly on tankers. He has taught satellite navigation, radar and ship handling part-time at a maritime college and run a three-bridge simulator.

128 | Also available

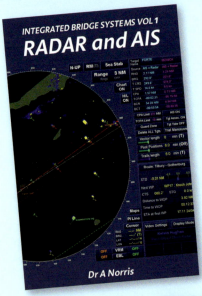
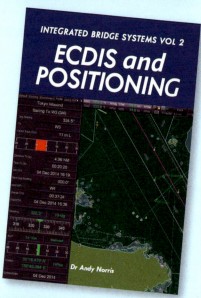

For more information on book discounts and membership please see our website
www.nautinst.org

THE NAUTICAL INSTITUTE